Democratization of Expertise
How Cognitive Systems Will Revolutionize Your Life

Ron Fulbright

Department of Informatics and Engineering Systems
University of South Carolina Upstate
Spartanburg, South Carolina, USA

CRC Press
Taylor & Francis Group
Boca Raton London New York

CRC Press is an imprint of the
Taylor & Francis Group, an **informa** business

A SCIENCE PUBLISHERS BOOK

CRC Press
Taylor & Francis Group
6000 Broken Sound Parkway NW, Suite 300
Boca Raton, FL 33487-2742

Version Date: 20200302

International Standard Book Number-13: 978-0-367-22964-1 (Hardback)

**Visit the Taylor & Francis Web site at
http://www.taylorandfrancis.com**

**and the CRC Press Web site at
http://www.crcpress.com**

Preface

We are standing at the edge of revolution of the human condition. A confluence of several technologies is bringing forth a new class of tool—cognitive systems—able to perform cognitive processing at, or exceeding, the level of a human expert. Over the last two hundred years, the industrial revolution, computer age, information age, Internet age, and the social media age have changed the way we live, work, and play several times over. The overall pattern in each of these ages is technology, once available only to a small set of specially trained or uniquely knowledgeable people, becomes available to the masses and the mass adoption of the technology changes everything.

In the coming cognitive systems era, humans and cogs will work together in natural, collegial partnership and collaboration where the total amount of cognition achieved by the human/cog ensemble is a combination of human and artificial thinking. When cogs become available to the mass market via Cloud, Internet, and the devices we use every day, billions of average humans will have the ability to perform at the level of an expert in virtually any domain—the *democratization of expertise*. This will lead to dramatic social, cultural, and economic changes as all revolutions do. The cogs are coming!

Preface

Acknowledgements

I would like to acknowledge the hard work, patience, and diligence of a number of my talented graduate students in the Master of Science in Informatics program and undergraduate students in the Information Management & Systems program at the University of South Carolina Upstate. In alphabetical order: Antonella Comancho, Kristen Good, Dane Jordan, Sergio Martinez, Victoria Mathis, Royston McKaig, Alantra Middleton, Karl E Riley, Ashley Rowland, Steven Stanton, Elizabeth Sullivan, and Grace Vaughan.

December 2019   Ron Fulbright

Contents

Chapter 1
When Things Become Democratized

When something becomes available to the average person, we say it has become *democratized*. Throughout history, technology and knowledge once accessible only to the elite in society has become available to the masses. Every time it happens, mass adoption of new technology brings about cultural, social, political, and legal changes. We are now entering the "cognitive systems era" in which access to cognitive systems technology will change everything!

The term *cognitive systems*, refers to technology making it possible for artificial constructs (computers, robots, and software) to perform high-level cognitive processing normally attributed to human thinking. This is not to say these cognitive systems are artificially intelligent. Some cognitive systems may be classified as being artificially intelligent but other cognitive systems are not intelligent, they just perform some cognitive processing at or above the human level.

We foresee a future, not too many years away, in which millions, even billions, of humans routinely work with and collaborate with cognitive systems technology on a daily basis. When they do, average humans will be able to perform at the level of a human expert in a particular domain of discourse—something we call *synthetic expertise*.

The mass adoption of cognitive systems and synthetic expertise will lead to changes all throughout our society culture. Recent decades have seen the adoption of personal computer technology change everything. Over the next few decades we predict the adoption of cognitive systems technology to change things just as much or more.

1.1 Adoption of Technology

The word *democratization* in this context does not refer to a political form of government or a type of voting. Rather, democratization refers to when a technology has become available to, and used by, the majority of the population (Friedman, 1999). Adoption of new technology takes some time. Everett Rogers described technology adoption in his book *Diffusions of Innovations*. The originators of the technology, the *innovators*, constitute the first users of a new technology but represent only a small portion (2.5% or 1 in 40) of the eventual market. The *early adopters*, those eager to use new technology, represent about 13.5% of the market. Early adopters await the announcement of a new technology and adopt it quickly even before the new technology is proven. Together, the earliest group of users represent about 16% or 1 in 6 of the market (Rogers, 2003).

Most people do not jump on new technology right away but instead come along at a later time. Being more cautious, the *early majority* users (34% or 1 in 3) learn about a new technology after it has gained some following and received positive feedback from the early adopters. By the time the early majority arrives about 50% of the eventual market has adopted the new technology and we say the technology has gone *mainstream*. Once a technology goes mainstream, the other half of the market learns about it and becomes convinced to adopt it.

The *late majority* users represent another third of the market. Once the late majority arrives, some 84% of the market has adopted the technology and we say the technology has become *dominant* representing the peak usage of the technology. After the peak has been reached, late adopters called *laggards*, will come to use the technology for the first time, but mainstream users will be falling away in favor of the next new technology. The adoption cycle is depicted in Fig. 1-1 (Gartner, 2019).

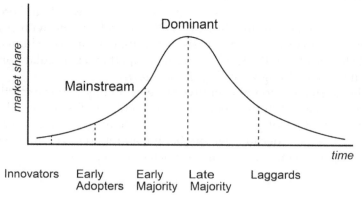

Fig. 1-1: Adoption of technology (following Rogers).

The amount of time required for a product or service to achieve dominance varies. Also, the length of time a product or service remains in a dominant position varies. However, there are rarely "overnight successes." In technology it typically takes a generation (20–25 years) for items in research and development labs to become mainstream products and services.

Eventually, users fall away as they move to the next generation of the product or a different product altogether. Joseph Schumpeter called this *creative destruction*. New creations take over making the old obsolete (Schumpeter, 1942). Generally, this can happen in two ways. Clayton Christensen describes *sustaining innovations* as incremental or evolutionary improvements to a product or service (Christensen, 1997). An example of these kinds of improvements are this year's model of smartphone. The new model has improved features but is not radically different from the previous model.

Christensen identifies *disruptive innovations* as those bringing about new consumer behavior, displacing existing markets, and creating new markets altogether. When the iPhone was released in 2007, it fundamentally changed the mobile, handheld communication industry. Previously, mobile phones had been more business-oriented but Apple's iPhone platform allowed developers to quickly deploy new applications allowing the iPhone to become a mass-market device for personal use. The social media, photography, and online shopping revolutions the iPhone kicked off are still in progress.

Today, we are quite used to and comfortable with technological change. In fact, we expect technology improvements every year. Peter Drucker discusses continual improvement as a response to opportunities in a market (Drucker, 2006). Producers recognize the needs and desires of consumers and respond by offering new and better products and service to meet those needs.

The central theme of this book is cognitive systems technology bringing forth many disruptive changes to how we live, work, and play. The technologies leading to the cognitive systems era have been in development for many years so we are well in to the research and development era of the technology. We foresee the cognitive systems mass-market adoption happening over the next decade.

1.2 Revolutions

We get an idea of what is about to happen by looking back at some historical revolutions brought about by mass-market adoption of new technology. Note the importance of getting a new technology to the point where millions, if not billions, use it. Mass-market adoption is key to culture-altering innovations.

The Printing Press

Until computers and the Internet came along, the printing press was the most influential information technology ever invented by humans (Eisenstein, 1983). Movable type was invented in China over two thousand years ago and was used in numerous civilizations. However, it wasn't until the 1400s when improvements by Johannes Gutenberg brought the printing press to the masses.

Gutenberg's hand-held mold made it possible for practitioners to quickly cast sets of type—formerly a very time-consuming task requiring specialized skill. In addition, Gutenberg used an oil-based ink better suited for printing on the paper in use at the time. Finally, Gutenberg's changes to the press itself made it easier and cheaper to quickly print multiple copies.

Printing had formerly been the purview of skilled craftsmen with each copy requiring significant time, effort, and special skill. In Europe at the time, copies were made by manually scribing the copy (manuscript) or block printing involving the carving of an entire page, one page at a time. Gutenberg's printing press made it possible for others to enter the printing industry. Becoming available at the beginning of the Renaissance, the printing press spread quickly through Italy and then through the rest of Europe and ultimately the entire Western world displacing manuscript and block printing. Never before had it been possible for a person to quickly and efficiently communicate to thousands, or tens of thousands, of people across great geographic distances. The era of mass communication had begun and changed everything. Societal, cultural, political, and religious revolutions followed.

An example of the effect of the printing press is the Reformation. In the 1500s, Martin Luther and others protested the authority of the Roman Catholic Church. The Reformation led to a revolution in religion, politics, and culture still reverberating throughout the world today. The Reformation would not have been possible without the ability to communicate across Europe via printed publications.

Another example of the transformative power of the printing press is education and access to knowledge. Before the printing press, knowledge largely remained within local communities. The ability to capture knowledge into books and spread that knowledge across vast distances transformed culture and fueled the Renaissance. For the first time, scholars could pass their knowledge on to other scholars who could build on ideas. Printing also made expert knowledge available to the average person. This both necessitated and also facilitated education of the masses. Whereas previously, education was available only to the elite in society, now everyone can be educated.

The printing press brought about the first democratization of information and knowledge. Not until the Internet has any other piece of new technology affected the lives of average people more than the printing press.

Electricity

The modern world has become dependent on electricity. People expect electricity to always be available and when it is not, due to a storm or other temporary outage, peoples' lives come to a halt. Yet just a little over a century ago, electricity was far from a guarantee. The electrification of the developed world has changed everything.

The principles of electromagnetic generation were discovered by Michael Faraday in the 1830s but it was not until the late 1800s when central power stations were built and electricity distributed to commercial customers. Household electrification in the United States and Europe began in the early 1900s and took another thirty years to reach 70% (moving past the "dominant" threshold of technological adoption). Even though it took a century, the democratization of electricity changed nearly everything about society and culture. The key was the building of the infrastructure to deliver electrical power to average people's homes.

Automobiles

Many people mistakenly believe Henry Ford invented the automobile. However, Ford made it possible for the masses to *afford* an automobile and that changed everything. Karl Benz invented the modern motorized vehicle in 1885 but it took another 23 years for Ford's Model T to become available at a price average people could afford. Until then, motoring was an expensive hobby only the rich could engage in. Designing the Model T to be easy and cheap to build using the assembly line concept allowed the mass production of cars and lowered the cost of each car. As a result, millions bought their first cars within only a few years of time.

Over the next thirty years following the introduction of the Model T, the automobile industry was created, the automobile went mainstream, and cultural/societal changes followed. Mass use of automobiles enabled people to move away from city centers creating the urban landscape. After an interruption because of World War II, Americans in the 1950s found themselves with both leisure time and money. Using their automobiles, the vacation industry was created.

Automobiles and the burgeoning national highway system enabled people to travel anywhere within the country. Although the Gold Rush of the mid-1800s brought the first wave to California and the west coast of the United States, the automobile brought millions in the 1950s and 1960s to California and the entire west coast. As a result, the landscape of the United States changed forever. American culture changed also and at least two generations grew up identifying with the automobile. Today, the automobile, like electricity, is assumed and expected.

Radio and Television

Although radio technology was invented in the 1880s and 1890s based on discoveries decades earlier, it was not until fifty years later, in the 1930s, radio became a mass-market success. Television followed a similar path taking several decades to go from basic scientific discovery to commercial success after World War II.

However, radio and television for the first time gave humanity the ability to converse in real time visually and audibly with millions of people. Previously, people received news and other information via printed newspapers, magazines, and books. The only audio-visual form of receiving news were the newsreels in movie theaters. Radio and television transformed the way people learned things from other parts of the world. Radio and television also transformed entertainment with books, live theater, movies, magazines, and records being the primary entertainment channels previously. As motion pictures had given rise to the motion picture star, radio and television created radio stars and television stars.

Radio and television intimately connected humans in ways print material could never do however the type of connection was one-way only. Consumers of content could not communicate in real time back with the producers of the content nor could consumers interact with the content. That changed in the computer/Internet revolution discussed next.

Personal Computers, the Internet, and Social Media

Advances in microelectronics and computers combined to form the personal computer market in the 1970s. Computers were invented in the 1940s and 1950s and continued to evolve throughout the 1960s as large, expensive tools for companies and government organizations. The invention of the microprocessor brought computers down in size and cost. These hardware improvements along with advancements in software resulted in the democratization of computers and information processing.

In the early days of computing, computers were owned and operated by only the largest companies and government agencies. Computers were accessible only to those with special knowledge and training. Desktop and personal computers running easy-to-use software brought the computer to the masses and has changed everything. We have now all become information workers and knowledge workers.

The Internet, the interconnection allowing bi-directional flow of data between computing devices, was invented in the late 1960s but reached mass-market significance in the 1990s only after being combined with the personal/desktop computer. However, one more advancement was needed to allow the average person to create content for the Internet. World Wide Web technologies democratized the Internet and allow billions of people to not only access the Internet but also create content for other people via the Internet.

Today, the biggest entities on the Internet including Google, Wikipedia, eBay, YouTube, Facebook, Twitter, Instagram, and Snapchat are based on *user-generated content (UGC)*. In the past, newspapers, magazines, book publishers, movie studios, radio stations, and television networks created the content and distributed it to the consumer. Today, we are squarely in the *social media era* in which the average person creates news and entertainment content. This UGC is distributed to billions of other average people via the Internet and consumed by them through their handheld computer devices (smartphones), tablets, and laptop computers.

1.3 The Nature of Revolutions

The democratization of the computer, the Internet, and content generation through the *computer age,* the *information age,* and the *social media age* has changed everything about how we live, work, and play. Radio, television, electricity, the automobile, and the printing press caused similar disruptions to how we entertain ourselves and communicate with others. These new technologies caused revolutions in how we think, what we value, and even what we believe.

The first thing to note about these kinds of revolutions is although it sometimes seems as if new technology springs out of nowhere, technological-fueled revolutions are actually the confluence of technology in a number of fields each with its own long history of progress and advancement. The printing press combined wine presses, ink, and moveable type all of which were in existence for a millennium before the invention of the printing press that changed everything. Ford's automobile combined industrial assembly-line manufacturing and gas-powered personal transportation devices both of which already existed. The social media age we are in now is the confluence of the computer, software, and microelectronics industries each having their own decades-long evolutionary trajectories.

The second thing to note about technological revolutions is it usually takes several decades for the individual components of the revolution to develop to a point where their convergence creates something revolutionary. Ford's automobile took on the order of 30 years after Benz's invention to reach market dominance and change culture and society. Personal and desktop computers reached market dominance 40 years after the invention of the computer. It took an additional 25–30 years for the social media era to evolve from the computer age.

1.4 The Cognitive Systems Revolution

We take note of the existing technological revolutions discussed in this chapter because we are at the beginning of a new technological revolution—the *cognitive systems era*. The remainder of this book discusses cognitive systems and the coming era in detail. Like other technological revolutions, cognitive systems are a confluence of several technologies including: artificial intelligence, machine learning, deep learning, big data, the Internet of Things, the Internet, cloud-based services and data, handheld devices, natural language interfaces, and open-source artificial intelligence.

Each of these technologies have a long history of evolution going back several decades. Individually, they can lead to significant goods, products, and services. However, together, they form a technological revolution promising to once again change everything about how we live, work, and play.

Chapter 2

The Coming Cognitive Augmentation Era

The theme of this book is how, in the coming years, mass adoption of cognitive systems technology will bring about cultural, social, and political changes. Research and development in cognitive systems is producing artificial systems, called *cogs*, able to perform high-level cognition. Until now, humans have had to do all of the thinking. However, soon, our computers, handheld devices, cars, and all other objects we interact with on a daily basis will be able to perform human-level thinking with us and without us. The future will belong to those better able to partner with and collaborate with these devices. Over the next fifty years, the infusion of intelligence into our lives will bring about as many changes as the computer/Internet/social media revolutions has over the last fifty years.

This is a bold prediction. Why should we believe such a prognostication? As described in Chapter 1, technological revolutions do not happen overnight. Rather, technological revolutions are the confluence of several different lines of technological development. The coming cognitive systems era is no different. The cognitive systems revolution is the convergence of the following:

- Deep Learning/Machine Learning
- Big Data
- Internet of Things
- Natural Language Interfaces
- Open-Source AI
- Cloud-Based Services
- Social Media
- The Connected Age

2.1 Deep Learning/Machine Learning

Programming a computer to learn has been the goal of artificial intelligence researchers since the early days of artificial intelligence as a discipline. Arthur Samuel coined the term *machine learning* in 1959 and helped develop some of the first computer programs that improved over time with experience (Samuel, 1959).

As described in more detail in Chapter 4, in the 1960s and 1970s, researchers first attempted to use symbolic reasoning to achieve machine learning. However, representing knowledge as a set of symbols is difficult due to the quantity of knowledge required for intelligence and the "fuzziness" (ambiguity) of knowledge itself. In the 1970s and 1980s, expert systems captured knowledge obtained from experts in the form of sets of production rules (if..then logic statements). However, these knowledge stores were not learned by the computer (Carbonell, 1983). Instead, humans captured and carefully engineered experts' knowledge and the nuances of that knowledge. The knowledge engineering effort required a large amount of time, effort and resources.

In the 1970s, a new way to do machine learning became prominent— neural networks. Inspired by the human brain, a neural network is a collection of highly-connected circuits each with a weighting factor determining the strength of the connection. A network is presented with a stimulus creating an array of signals throughout the network. Weights are adjusted to get the network's response closer to the ideal and then it is presented with another stimulus. Over time, the weights associated with the connections are tweaked until the network's response to a class of stimuli is firm. When the network is presented with a slightly different stimulus from the ones used in the training set, it is still able to recognize it as belonging to a certain class of stimuli. This makes neural networks quite robust in the face of incomplete data (which the real world is composed of).

However, like with expert systems, training neural networks required large amounts of careful knowledge engineering to select positive and negative training examples and to provide feedback to the network. As a result, by the 1990s, machine learning research by any method had largely been abandoned.

Advances in computer speeds, reduction in costs, availability of enormous datasets, and new statistical-based algorithms led to a resurgence of interest in machine learning during the 1990s and into the 2000s. With much lower barriers to entry and fueled by the "dot com boom" of the late 1990s, many startup companies emerged tackling tasks such as natural language understanding, computer vision, image classification, handwriting recognition, and data analytics.

Deep learning employs multiple layers of neural networks allowing the extraction of higher-level features from a stimulus (instead of just one response from a single-layer network). Allowing a computer to learn multiple things at different levels of abstraction led to the deep learning revolution. By 2011 and 2012, deep learning systems were demonstrating superhuman performance in handwriting recognition and image classification. In 2012, a deep neural network named AlexNet won the annual ImageNet Challenge beating all competitors by a substantial margin (Krizhevsky et al., June 2017). The success of this deep neural network garnered interest not only within the artificial intelligence community but across the entire technology industry.

A startup called DeepMind began in 2010. Using convolutional deep neural networks and a proprietary machine learning algorithm called Q-Learning (Watkins, 1989). DeepMind built computer systems able to learn how to play 1980s-style video games without training sets. The system learned how to play simply by watching video games being played. This is called *unsupervised learning* and is an important component of deep learning and the cognitive systems revolution. With unsupervised learning and self-supervised learning, there is no need for time and effort-consuming knowledge engineering like there was in previous machine learning eras. This opens the horizon.

Google purchased DeepMind in 2014. In 2016, a system called AlphaGo defeated the reigning world champion in Go, a game vastly more complex than Chess. In 2017, a version called AlphaGo Zero learned how to play Go by playing games with itself and not relying on any data from human games. AlphaGo Zero exceeded the capabilities of AlphaGo in only three days. Also in 2017, a generalized version of the learning algorithm called AlphaZero was developed capable of learning any game. AlphaZero achieved expert-level performance in the games of Chess, Go, and Shogi after only a few hours of unsupervised self-learning (DeepMind, 2018a; DeepMind, 2018b; ChessBase, 2018).

Using unsupervised deep learning, future cognitive systems will be able to learn on their own and acquire expert-level knowledge and capability in periods of time that will astound us. These systems will not be programmed to perform in a certain domain, they will acquire that capability on their own and in their own way. Still in its early stages of development, the importance of unsupervised learning cannot be understated. Along with the other technologies described in this chapter, unsupervised learning will lead to systems able to learn and perform at expert levels in virtually any domain. This will drive the cost of producing such systems down so as to make them affordable to the masses leading to the democratization of expertise.

2.2 Big Data

If cognitive systems of the future are going to be consuming information and learning on their own, consideration needs to be given as to where this information will come from. Recently, advances in handling and processing enormous of unstructured data of various types has led to the big data industry. Big data can be characterized by the following (Jain, 2016):

Volume	Amount of data (exabyte and petabyte-sized stores)
Variety	Type of data (text, images, audio, video, etc.)
Velocity	Speed of data (generation, acquisition, processing)
Veracity	Quality of data (accuracy, reliability, validity, etc.)
Value	Worth of data (return on investment)

The field of big data involves ways to store, process, and analyze data stores too large and too unstructured to be handled with traditional database technology. The goal is to extract meaningful and useful knowledge from these vast data stores. Often, big data analytics exposes patterns and connections in the data humans have no reason to expect are there.

There are many examples of big data analytics. For example, Walmart handles more than one million customer transactions per hour and their analytics processes databases measured in exabytes. Walmart is continuously analyzing buying behavior so it can accurately configure its stores, and the products offered in their stores, to meet the demands of the customers near the store.

The Internet and social media provide examples of using big data. For example, NetBase analyzes people's postings on various social media platforms to determine people's sentiment on a topic (www.netbaase.com). Better than knowing a customer bought your product is to know why the customer bought the product and what they and others feel about your product.

Cognitive systems are currently looking at various big data stores and detecting patterns and relationships in the data. Through unsupervised learning as discussed in the previous section, cognitive systems will obtain knowledge on their own from these sources.

2.3 Internet of Things

Today, over four billion people use the Internet. Internet usage, especially social media and video, have exponentially increased the number of bytes humans generate each year. Humans are currently generating hundreds of exabytes (billions of gigabytes) per year and the trend will continue into the foreseeable future with the total amount of data generated doubling within two years and, within a few years, doubling every year. This alone gives big data systems zettabytes of data to process and analyze. However, another technology is coming promising to increase data generation even more—the *Internet of things*.

Internet of Things (IoT), also *ubiquitous computing*, refers to Internet connectivity being extended to devices and everyday objects such as: appliances, automobiles, bedrooms, toilets, etc. (Weiser, 1991). Making everything in our lives Internet addressable enables new kinds of applications and technology. It also promises to add even more data to the amount already generated by humans.

However, there is a difference. IoT data will capture much more meaningful data about us and our behavior. When a person watches a movie streamed from one of the Internet streaming services, the action generates on the order of a couple of gigabytes of data. How much can be learned about the person who is watching the movie? Certainly, a number of things can be learned from the time of day, day of the week, movie genre, etc. In fact, services use these data now to better serve their clientele. However, consider the difference between those two gigabytes of data and two gigabytes of data from the person's toilet and bedroom. These kinds of data open up new possibilities of learning about a person. By interacting with our devices in everyday life, we put out enough biomedical information to equal or exceed that of an office visit with the doctor.

The cognitive systems era will see the development of systems able to monitor the continuous stream of data we exude and detect things about us using unsupervised deep learning. The IoT will bring cognitive systems into intimate contact with us.

2.4 Natural Language Interfaces

Another key technology making the cognitive systems era different is Natural Language Interfaces (NLI). NLI refers to systems able to communicate with humans using the same spoken, written, and gestural language we use to communicate to other humans. In the cognitive systems era, devices and applications capable of high-level cognition will become ubiquitous. It will not be possible for humans to learn a special mode of communication with each and every device and application. These devices and applications will have to be able to communicate with humans in a natural way.

We are seeing the beginning of this now with our current voice-activated devices. People commonly speak to their smartphones now to issue commands and requests. Voice-controlled devices are beginning to enter and spread through the home as well. In the cognitive systems era, people will expect to be able to communicate with anything and everything via natural language.

Natural language understanding was once a major goal of artificial intelligence research. The first systems able to understand natural language were STUDENT and ELIZA in the mid-1960s (Weizenbaum, 1966). Siri, the voice-activated intelligent assistant on Apple iPhones, was introduced as a standard feature in 2011. This was the first voice-activated interface a large number of people interacted with even though voice control had been available on desktop and laptop computers for many years. Soon after, most smartphones in the market featured natural language voice control.

With hundreds of millions of people using them in the early 2010s, the ubiquitous nature of the smartphone brought natural language interfaces to the masses. In 2011, IBM's Watson demonstrated the ability to understand written natural language in the form of *Jeopardy!* "clues" and was able to formulate and deliver spoken natural language "answers" (Jackson, 2011).

Even though tremendous advances have been made recently, there is still a long way to go in NLI. Currently, it is not possible to carry on a casual conversation with an NLI device or application. Human language is amazingly complex and nuanced with notable differences even within different areas of a country. Research is proceeding rapidly in this area however (Alexa Prize, 2019). Conversational NLI is a requirement for the cognitive systems era. Humans will expect to be able to carry on conversations with cogs like they do with other humans.

2.5 Open Source AI

Throughout the history of technology, it has been demonstrated new and interesting things happen when technology becomes available to the average person. For example, in the 1970s, microprocessors and microelectronics became widely available and affordable to young enthusiasts like Steve Jobs and Steve Wozniak enabling them to create the personal computer industry with the introduction of the Apple I and Apple II computer.

Likewise, many progenitors of the "dot com" revolution in the 1990s began by experimenting with hand-me-down minicomputers and microcomputers in the 1970s and 1980s. Formerly available only to large companies and government agencies, electronics and computers started as playthings for a new generation of technology enthusiasts. They created the personal computer industry and the social media industry which now drives the entire technology sector.

The same has been true in the software industry. In the 1950s and 1960s, programming a mainframe computer was the purview of specially trained people. This form of software was not accessible to the average person. First of all, the computers running this software was not available to the masses and secondly, the technology of writing software was not suited for consumption by a large number of people.

From the beginning of computing, a few engineers shared software with each other to evolve the field. However, with the influx of personal and desktop computers in the late 1970s and through the 1980s, millions of people became interested in writing software for their machines. With personal computers becoming a mass-market item, also came the mass-market software industry. Bill Gates and Paul Allen formed Microsoft to supply software for this emerging mass market.

Along with the mass-market software industry, also evolved was the mass-sharing software community. Among software developers, there has always been the ethos of free software. This community has always felt software should be freely open to everyone with each person able to download, use, and even alter source code. The Free Software Foundation was opened in 1985 (www.fsf.org) and the Open Source Foundation was established in 1998 (www.opensource.com). However, by this time, the software industry had seen the open sharing of software in the 1970s and 1980s. In fact, some of the most important software running in the world today, such as the Apache Web server, the Mozilla Web browser, and the Linux operating system was created as open-sourced software.

Naturally, cognitive systems will be the result of artificial intelligence and deep learning software development. The open source ethos is spreading to this domain also. The OpenAI organization was created in 2015 (www.openai.com). OpenAI's purpose is to promote free and open exchange of artificial intelligence software. The OpenCog project started in 2008 and is an open-sourced artificial intelligence framework. Originally developed by Google, TensorFlow was released in 2015 as an open-sourced machine learning framework. In 2016, the Microsoft Cognitive Toolkit was released as a deep learning framework.

These, and others resources are available via the Internet. This gives excellent resources to millions of average people (technology/AI enthusiasts for sure). Major breakthroughs are sure to come from people all over the world tinkering around with these resources. Recently, the model has been a few young people experimenting with a new idea create a small startup company. Startups achieving even modest success attracts attention from the large companies such as Apple, Google, Facebook, Microsoft, and IBM. The most promising of the technology in these startups are acquired by the large entity when it buys the startup. The technology of the cognitive system era is likely to happen this way over the next several years as artificial intelligence and cognitive systems technology development by the masses explodes.

2.6 Cloud-Based Services

Of similar importance to open-sourcing is the widespread use of *cloud services*. The word "cloud" comes from the use of an amorphous region drawn on diagrams when one does not know or care of the implementation details within the part of the system contained in the "cloud."

Today, the *cloud* refers to on-demand computing, storage, and services available primarily through the Internet. Instead of maintaining expensive computer infrastructure internally, businesses can purchase access to such infrastructure. The actual equipment is located somewhere but is accessible as if it were in the next room. This dramatically lowers the cost of developing computationally expensive applications like artificial intelligence and cognitive systems.

The cloud is available to companies and also to individuals. Examples include Amazon Web Services (aws.amazon.com), Google Cloud Platform (cloud.google.com), and Microsoft Azure Cloud Computing (azure.microsoft.com).

Also available via the cloud is an enormous amount of data and services. A young start-up developer then has open-source software, data, services, and computing infrastructure available. Once something new is created, the cloud gives a new startup a way to offer its new services to customers. There may be millions of people out there on the Internet waiting for the new creation. Cloud-based development and cloud-based delivery makes the world today a very different business world from a few years ago. Previously, marketing and distribution consumed as much or more effort as creation of a new product. New products and services to be developed in the cognitive systems era will "live on the Internet."

2.7 Social Media

We are living in the social media age. Social media involves computer-mediated communication and sharing among millions (and sometimes billions) of geographically separated people and includes platforms like Facebook, Twitter, Instagram, Snapchat, YouTube, and LinkedIn.

Social media platforms enable participants to share texts, documents, photos, videos, and messages and as such is a user-generated content medium. Social media has changed the way people receive news, express opinions, interact with each other, and even has affected what people believe ethically and morally. This is not unexpected. As discussed elsewhere in this book, mass-market adoption of new technology often reverberates throughout cultural and societal behavior in this way.

The social media era, and in general, the "connected age" discussed in the next section, is a perfect environment for the introduction and mass adoption of cognitive systems technology. Billions of people around the world are primed for rapid adoption. Today, it is not uncommon for a new game, video, or posting to "go viral" meaning awareness of the item spreads quickly through person-to-person communication. Never before in the history of the world has it been possible for the average person to create something seen or heard by millions of people in only a few hours' time (or a few minutes in some cases). Apps and services in the cognitive systems era will be used by the average person and using them will achieve amazing results. People will naturally notify their friends via social media of the great new thing they have discovered. Celebrities will use cognitive systems apps and talk about them on social media only serving to increase the awareness and make adoption of the new technology happen at a very rapid pace.

2.8 The Connected Age

The enormous number of people already connected by, and comfortable using, high-bandwidth handheld electronic devices enables a low cost of cognitive systems technology via the Internet. Almost one-half of the world's population are already "connected," most of those engage in social media, and the majority of them engage in streaming and/or use cloud-based services. This describes a market ready for rapid adoption.

Most younger people have grown up relying on Internet-based resources and get the majority of their information and news from the Internet and social media. When cognitive systems technology shows up on the Internet in the form of apps and services, this generation of people will likely be the early adopters. They are so used to adopting new technology in this way they may not realize they are fueling a new revolution. They may just view cognitive apps as just being the expected "next great thing" they are used to seeing.

2.9 The Cognitive Systems Era

Even though each of the above-mentioned is the culmination of decades of development, they are just now coming together in a way suited for mass-market products. In short, it is the right time for something revolutionary. One revolutionary type of cognitive system technology will be in the *cognitive augmentation* arena—the enhancement of human cognitive ability. Recent successes have demonstrated systems utilizing one or more of the above factors capable of performing at the level of a human expert. Major companies such as IBM, Microsoft, Google, and Amazon are investing billions of dollars into developing "AIs." While these AIs are not personal cogs, we foresee the personalization of these kinds of systems as being the next step. When that happens, expert-level performance in a variety of domains will be in the hands of the ordinary individual.

We think personal cogs will redefine the relationship between people, their environment and their technology. Personal cogs will be available to us at any time of the day through a variety of interface methods, including natural language. We will come to view personal cogs as helpful entities always available in our environment whether at work, play, or leisure. Cogs will play the role of assistant or coach, will act autonomously and semi-autonomously, be prescriptive, suggestive, instructive, or simply entertaining.

The Cog Market

Large-scale cognitive systems we call *enterprise cogs* are being developed now. This will certainly continue and spread into every industry and will certainly become an economic force in the market worth many billions of dollars. However, enterprise cogs are not intended for purchase and use by the average person. We predict the emergence of *specialty cogs* and *personal cogs* intended for the mass-market. These cogs will be bought and sold by average people through existing sales channels much in the same way apps, music, and books are sold now. We will both be able to purchase or rent our own cogs and also be able to subscribe to Cloud-based cog services and information/knowledge services.

Because cogs work in partnership with humans, there will arise the need for experts in a field to work with cogs and develop their own unique store of knowledge (we call this a *cogbase*). Entities in industries such as financial services, investment services, legal, medical, news, politics, and technology will compete in offering access to their "superior" store of knowledge created through the interaction of their experts and their cogs. In the cog era, we predict the commoditization of expertise.

Teacher Cogs

We expect cogs to become a new form of intelligent tutoring systems. A teacher cog will have access to everything knowable about a particular subject matter. Through customized and personalized interaction with a person, teacher cogs will impart this knowledge to the student in ways similar to the master/apprentice model of education. The best teacher cogs will be personal cogs able to remember every interaction with a person over an extended period of time, even years or decades. Imagine an algebra cog able to answer a question by a 35-year old who it has been working with since grade school. We anticipate teacher cogs to evolve for every subject taught in schools and beyond. There may even be *master cogs* developed incorporating the domains of several individual teacher cogs. One can imagine an engineering cog being comprised of an applied calculus cog, an algebra cog, and a differential equations cog. We think students of future generations will start using cogs all throughout their education and then retain the cogs, and years of interaction, through the rest of their lives. Again, we foresee vigorous competition arising from different teacher cog providers attempting to bring to the market the best teacher cog for a particular subject matter.

Advisor, Coach, Self-Help, and Pet Cogs

Humans will interact with cogs using natural language. People will have conversations with cogs and the cogs will respond in creative, knowledgeable, and personalized ways. It is natural for humans to form emotional relationships with anything, biological or artificial, they can interact with. Indeed, people form emotional relationships with animals and technology today. We foresee cog technology giving personalities to artificial systems. Since cogs will be able to give expert-level advice in any domain, we predict the evolution of a host of self-help cogs ranging from relationship advice to life/work balance, to grief counseling, faith-based counseling and beyond. People will confide intimate details to these cogs and receive advice of great personal value and satisfaction. People will spend hours conversing with their personal self-help/companionship cogs. We can easily envision the development of virtual pets with cog-based personalities and communication abilities. In the cog era, we will love our cog pets.

Productivity Cogs

We predict every productivity application in use today will become enhanced by cog technology in the future. Indeed, applications like word processors, spreadsheets, presentation editors, Web browsers, entertainment apps, games, graphics editing, etc. may become a primary interface point for humans and cogs. Cog capabilities will both be built into the applications themselves and provide expert-level collaboration to the user and also evolve into stand-alone cogs for a particular task. For example, we can imagine a future version of Microsoft Word coming complete with embedded creative writing cog services. We can also imagine purchasing a creative writing cog from an app store operating independently of a specific word processor.

Personal productivity cogs will understand our recent context in a deep manner, like a spouse, coworker, or a friend, and use that to customize their assistance and interaction with us. Imagine, for example, a word processing cog that understands you are writing about the future of cognitive processing but also knows that you have communicated with several others via email on that and related topics and can also take into consideration every article or Web page you have accessed in recent months while researching the paper. Such a cog knows a lot about you personally and can combine that knowledge with its own searching and reasoning about the millions of documents it has searched on the Internet. Personal productivity cogs will become our intelligent assistants.

Collaborative Cognition

In addition to enhancing current productivity applications, we expect an entirely new genre of cog-based productivity app to arise, *collaborative cognition*. We envision new kinds of problem solving, brainstorming, business/competitive/market analysis, and big data analysis. We foresee multi-cog "collaborative virtual team" applications being created such as in Fig. 2-1. Such *cogteams* will consist of several cogs, each with their own domain of expertise, engaged in discourse with one or more humans and offering advice, answering questions, and performing research and analysis as the meeting dictates. Collaborative cogs will become our artificial intelligent team members. Again, we see a vigorous and dynamic competitive market arising around the idea of collaborative cogs. By partnering with humans, cogs achieve ever-increasing levels of knowledge in a particular area. Therefore, considerable market value will be attached to collaborative cogs that have worked with the best experts in the field. The cog era will bring forth a new kind of virtual consultant.

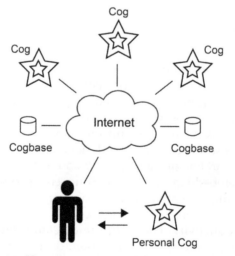

Fig. 2-1: Human partnering with local and remote cogs.

Research Cogs

We foresee future graduate students, entrepreneurs, scientists and any of us creative and inquisitive people conducting research by conversing with their research cog(s) instead of searching and reading scores of journal articles and technical papers. Today, I tell my graduate students the first step in their research is to go out and read as many articles, books, and papers as they can find about their topic and I try to give them guidance. My future research students' first action will be to sit down with his or her research cog and ask "So what is the current state of the art in <insert domain here>." The cog will have already consumed all existing material about the subject, will have recognized patterns and trends, and will be able to answer questions and even suggest lines of thought.

Cogs will be able to consume billions of articles, papers, books, Web pages, emails, text messages, and videos. This far exceeds the ability of any human. Even a person spending all of their professional life learning and researching a particular subject is not able to read and understand everything available about that subject. Yet, future researchers will be able to *start* their education from that vantage point by the use of research cogs. In the cog era, the best new insights and discoveries will come from the interaction between researchers and their research cogs.

Here again we see evidence of knowledge becoming a commodity. Today, we may be able to learn a great deal from the notebooks of great inventors like Tesla, Edison, and DaVinci. In fact, notebooks of inventors like these are worth millions of dollars. But imagine how valuable it would be if we had access to Einstein's personal research cog he used for years while he was synthesizing the theory of relativity? In the cog era, not only will cogs assist us in coming up with great discoveries, they will also record and preserve that interaction for future generations. Such cogs will be enormously valuable both economically and socially. See Chapter 11 for details on *synthetic colleagues*.

Discovery Engines

Even though cogs are intended to partner with humans and improve their knowledge and ability over time as a result of this interaction with humans, cogs will evolve to be able to perform an enormous amount of cognitive work on their own. We fully expect cogs working semi-autonomously and even autonomously to discover significant new theories, laws, proofs, associations, correlations, etc.

In the cog era, the cumulative knowledge of the human race will increase by the combined effort of millions of cogs all over the world. In fact, we foresee an explosion of knowledge, an exponential growth, when cogs begin working with the knowledge generated by other cogs. This kind of cognitive work can proceed without the intervention of a human and therefore proceed at a dramatically accelerated rate. We can easily foresee the point in time where production of new knowledge by cogs exceeds, forever, the production of new knowledge by humans.

In fact, we anticipate a class of *discovery engine cogs* whose sole purpose is to reason about enormous stores of knowledge and continuously generate new knowledge of ever-increasing value resulting ultimately in new discoveries that would have never been discovered by humans or, at the very least, taken humans hundreds if not thousands of years to discover. See Chapter 12 for details on automated knowledge discovery.

Cognitive Property Rights

Today, intellectual property rights represent a significant value, as much as a third of the US gross domestic product. The cog era will bring forth new questions, challenges, and opportunities in intellectual property rights. For example, if a discovery cog makes an important new discovery, who owns the intellectual property rights to that discovery? An easy answer might be "whoever owned the cog." But, as we have described, we anticipate cogs conferring with other cogs and using knowledge generated by other cogs. So a cog's work and results are far from being in isolation. We predict existing patent, copyright, trademark, and service mark laws will have to be extended to accommodate the explosion of knowledge in the cog era.

Synthetic Expertise

What will the world be like when most of us can perform as experts in any field? We are at the very beginning of the cog era and its evolution will play out over the next few decades. The cog era will, for the first time, give humans artificial systems to assist them in thinking. This cognitive augmentation will create a new industry and market and will change us culturally, legally, socially, and politically. If we let history be our teacher, we can expect the cog era to bring about changes we can only begin to imagine today. We now think of a time before electricity or a time before the telephone or television, or even before the computer and the Internet. We believe one day we will look back and think of a time before cogs. It's difficult to remember or imagine life without our modern technological conveniences and we believe one day we will say the same thing about cogs. This section contains some, but certainly not all, of the interesting factors that will come in the cog era.

Chapter 3
A Brief History of Human Augmentation

One thing setting humanity apart from all other animals on Earth is the ability to create new technology: tools, fire, wheel, language, mathematics, printing, computers, television, radio, computers, the Internet, social media, etc. Technology augments human ability and helps individuals do things they could not do before. This is turn changes how humans live, work, play, and evolve. Humanity creates technology and technology changes the trajectory of human history. The subject of this book is how, in the coming years, cognitive system technology will change everything and make it possible for average humans to do things we can barely envision today.

3.1 Technology Creates New Kinds of Workers

As technology and knowledge evolves, so does the nature of the *worker* (Fulbright, 2016a; 2016b). Long ago, humans scavenged for food. New technology such as bows, arrows, and spears allowed humans to hunt for food rather than scavenge. A new kind of worker was created—the *hunter*. Hunting changed the nature of humanity. Not only did hunting provide a reliable high-calorie, high-protein food source, society and culture organized around the idea of nomadic tribes following herds of animals. Everything else in these societies was built around the hunting culture.

Later, new technology enabled humans to farm animals for food and new workers called *ranchers* were created. Likewise, technology for growing plant-based food sources gave rise to the agrarian worker, creating *farmers*. Ranchers and farmers gave humans the ability to grow food where they lived giving rise to the establishment of new orders—towns and cities. Without having to travel around continuously following game, humans could live in one place. This allowed more people to congregate than was possible before in nomadic tribes, facilitating new kinds of communication between individuals, and necessitated another invention—*government*—eventually leading to nation-states.

Harnessing various kinds of energy has been key to numerous technological innovations. One of the earliest forms of harnessed energy was fire. In addition to the ability to cook food, fire enabled metalworking leading to the *bronze age* and the *iron age*, new kinds of metal tools and alloys and new workers—*iron workers* and *metal smiths*. Metalworking swept across the world thousands of years ago and transformed humanity—especially warfare—giving rise to kingdoms and dynasties.

Also of ancient origin is sailing technology harnessing wind energy and producing workers known as *sailors*. Sailing allowed humans to travel great distances across oceans and seas. Naval technology transformed the entire planet giving rise to global empires such as the British Empire. Sailing also enabled the European move westward and the colonization of the New World. Almost everything in global politics and society today has ties back to the colonial period facilitated by sailing and naval technology.

Later, the harnessing of steam, coal, oil, and electrical energy fueled the industrial revolution of the 1800s and 1900s and produced a new kind of worker—the *factory worker*. There has never been a greater agent of change than the industrial revolution. Almost everything in daily life has been affected directly or indirectly by the industrial revolution. Life became centered around the factory and people began organizing in self-sustaining communities adjacent to the factory called *mill hills* in the textile industry. In these communities, the company owning the factory provided not only the source of income but also ran the General Store, school, and medical services—essentially becoming small towns within larger cities.

The manufacturing industry, and the increasingly global marketplace, created the need for a new kind of worker to handle the information associated with the manufacturing business—the *office worker*. In the first half of the 1900s, many machines were invented to help office workers perform tasks. In fact, International Business Machines (IBM) made billions of dollars making things like tabulating machines, time clocks, cash registers, and later, computers for the business world.

Electronic computers ushered in the *computer age* and began a series of transformations which are continuing today. In the 1950s and 1960s, large "mainframe" computers transformed how companies did business. At the time, computers were affordable only by the most affluent companies and government organizations. When computers became desktop fixtures during the 1980s, they changed the nature of the office worker. Since then, office workers have become *information workers*. Today, thanks to the infusion of information technology, the average office worker's predominant effort is to process electronic information. We are still living in the *information age*.

The democratization of information processing has transformed the United States over the last 50 years in a significant way. The industrial revolution transformed the United States into a manufacturing-based economy where the item of value is what you make. Since the desktop and personal computer revolutions began, the United States has become a services-based economy where the item of value is what you know (the information and knowledge).

In addition to information workers, managing knowledge has produced the *knowledge worker*—one who exploits information—leading to disciplines like business intelligence and competitive intelligence, data analytics, etc.

What is beyond knowledge workers? What new kinds of roles and workers will future technology create? Expertise and wisdom is beyond the knowledge level and the future technology designed to manipulate expertise is cognitive systems—systems able to perform high-level cognition. This technology will allow anyone to perform at the level of an *expert* in a given domain. The new kind of worker this technology will create is the *synthetic expert*. Synthetic expertise will be available to everyone, something we call the *democratization of expertise* (Fulbright and Walters, 2020). Humans partnering with systems able to perform some of the thinking on its own will augment human cognition and lead to changes throughout society and culture.

3.2 Human Cognitive Augmentation

For millennia, tools have augmented the physical capabilities of humans. In the computer and information ages, tools have augmented the cognitive ability of humans—the ability to think. The idea of enhancing human cognitive ability with artificial systems is not new. In the 1640s, mathematician Blaise Pascal created a mechanical calculator called the Pascaline (Chapman, 1942). Thousands of years before this, mechanical devices such as the abacus aided basic arithmetic operations. Throughout history, humans have created many such devices. Using these devices, a human can perform mathematical calculations difficult or impossible for an unaided human. These devices augment human mental performance, but the human still does all the thinking.

In the 1840s, Ada Lovelace was among the first to envision a machine performing a human task—musical composition (Hooper, 2012; Isaacson, 2014). Lovelace imagined the machine composing the music, not a machine enhancing a human's ability to compose music. However, ideas like this were a century before their time. In the 1940s, Vannevar Bush envisioned a system called the Memex and discussed how employing associative linking could enhance a human's ability to store and retrieve information (Bush, 1945). Interestingly, associative linking through hyperlinks is at the very heart of the Internet and the World Wide Web technology. Like the above-mentioned calculating devices, the Memex made the human more efficient but did not actually do any of the thinking on its own.

Turing (1950) discussed if machines themselves could think and offered the "imitation game" to decide if a machine is exhibiting intelligence. Since coining the phrase *artificial intelligence* (AI) in 1955, at least three generations of researchers have sought to create an artificial system capable of human-like intelligence (McCarthy et al., 1955). Expert systems using if-then production rules have captured experts' knowledge and simulated the logical inference of experts in a field of discipline (Hayes-Roth et al., 1983). But these systems require laborious programming and are brittle in the face of uncertain or conflicting information. Artificial neural networks (ANNs) have demonstrated a human brain-like response to multiple, partial, and uncertain stimuli including adaptability and learning. But these systems lack general strategic and tactical reasoning ability such as goal-based search, planning, and synthesis. Chapter 4 contains a more detailed history of artificial intelligence.

Ross Ashby coined the term *intelligence amplification* maintaining human intelligence could be synthetically enhanced by increasing the human's ability to make appropriate selections on a persistent basis (Ashby, 1956). But again, the human does all of the thinking. The synthetic aids just make the human more efficient.

In the early 1960s, Engelbart and Licklider envisioned human/ computer symbiosis. Licklider imagined humans and computers becoming mutually interdependent, each complementing the other (Licklider, 1960). However, Licklider envisioned the artificial aids merely assisting with the preparation work leading up to the actual thinking which the human would do. Engelbart's H-LAM/T framework described the human as a part in a multicomponent human/computer system allowing human and artificial systems to work together to perform problem-solving tasks (Engelbart, 1962). Through the work of Engelbart's Augmentation Research Center, and other groups in the 1950s and 1960s, many of the devices we take for granted today were invented as "augmentation" tools including: the mouse, interactive graphical displays, keyboards, trackballs, WYSIWYG software, email, word processing, and the Internet. However, while making it easier for the human to think and perform, none actually do any of the thinking themselves. The idea of human and machine working together is a powerful one though and one cogs are bringing into reality.

As shown in Fig. 3-1, Engelbart's framework envisions a human interacting with an artificial entity while working together on a task.

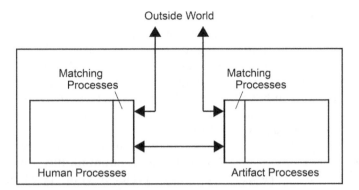

Fig. 3-1: Engelbart's H-LAM/T framework.

To perform the task, the system executes a series of processes, some performed by the human (explicit-human processes), others performed by artificial means (explicit-artifact processes), and still others performed by a combination of human and machine (composite processes).

According to Engelbart, augmentation can be accomplished by making improvements to any part of the H-LAM/T framework. For example, improving the language or training augments the performance of the human/artifact combination.

Being conceived in the 1960s, Engelbart's artifacts were never envisioned to do any of the high-level thinking themselves. In the coming cognitive systems era, the *artifacts* themselves are about to change. Recent advances in machine learning and artificial intelligence research indicates the artifacts are quickly becoming able to perform human-like cognitive processing. We call systems able of performing high-level cognitive processing *cogs*. We later update Engelbart's framework to create the human/cog ensemble wherein some high-level cognition is done by the human and some is done by the cog and the result of the ensemble is an emergent product of both kinds of cognition.

The concept of machine learning has long been a staple of artificial intelligence research. However, most of the early machine learning required careful and precise engineering and preparation of the information prior to the introduction to the machine doing the learning. We desire systems able to learn in the real world and with unstructured data and information. We are beginning to see this come true with *deep learning* (Deng and Yu, 2014; Itamar et al., 2013; Krizhevsky et al., June 2017). Recent advances in deep learning have resulted in systems learning multiple levels of representations corresponding to different concepts at different levels of abstraction. The levels form a hierarchy of concepts and result in deep understanding. Furthermore, deep learning algorithms learn from unstructured, voluminous, and disparate forms of information such as by watching videos, consuming millions of documents, mining social media, and observing human behavior.

Recent advances have brought to the realm of possibility systems capable of performing cognition on their own. One branch of AI has sought to develop semi-autonomous *intelligent software agents* to act on behalf of a user or other program (Nwana, 1996; Schermer, 2007). These agents are designed to interact as if they were human but also perform on their own without supervision from the human user.

The concept of the agent—a self-contained, interactive and concurrently executing object, possessing internal state and communication capability—can be traced to the Hewitt's Actor model (Hewitt et al., 1973). Software agents act autonomously and only occasionally communicate with the human user. However, the field of *human-autonomy teaming* has studied real-time interaction between humans and artificial systems. One active area involves military applications. Constraints must be met though because combat requires systems to respond rapidly and efficiently while attaining mission objectives (Barnes et al., 2017). One goal of this research is an Autonomous Squad Member where a human squad member, either in a military or law enforcement setting, is assisted by an autonomous agent in mission environments (Chen et al., 2017). NASA has researched the idea of having fewer human operators on long space flights by using artificial intelligence (Shivley et al., 2016; 2018). However, these are all highly-specialized applications. What about the average person?

More recently, Forbus and Hinrichs (2006) described *companion cognitive systems* as software collaborators helping their users work through complex arguments, automatically retrieving relevant precedents, providing cautions and counter-indications as well as supporting evidence. Companions assimilate new information, generate and maintain scenarios and predictions, and continually adapt and learn, about the domains they are working in, their users, and themselves. Like companions, the personal cogs we envision will operate in a limited domain, however, will achieve expert-level performance in that domain, often exceeding that of a human expert. Like truly human companions, personal cogs will be able to interact through spoken natural language, although not be limited to this form of communication.

Langley (2013) challenged the cognitive systems research community to develop a synthetic entertainer, a synthetic attorney, and a synthetic politician as a way to drive future research on integrated cognitive systems. The vision here is to develop a virtual human. We maintain the goal should be not to create a virtual human capable of being an entertainer, an attorney, or a politician, but rather create a cognitive system capable of expert-level performance in entertainment, a different cognitive system capable of exhibiting expert performance in a subfield of law, and a cognitive system capable of expert politicking. This is indeed the vision of IBM as it commercializes its Watson technology. We feel the natural extension of this technology will result in personal cogs being virtual entities capable of expert-level collaboration in a relatively narrow domain of discourse. Collaboration with the personal cog will enhance the human user's cognitive ability.

A significant step toward visions like these occurred in 2011, when a cognitive computing system built by IBM, called Watson, defeated two of the most successful human *Jeopardy!* champions of all time (Jackson, 2011). Watson received clues in written natural language and gave answers in natural spoken language. Watson's answers were the result of searching and deeply reasoning about millions of pieces of information and aggregation of partial results with confidence ratios. Watson was not programmed to play *Jeopardy!* Instead, Watson was programmed to *learn* how to play *Jeopardy!* which it did in many training games with live human players before the match (Ferrucci et al., 2010; Ferrucci, 2012). Watson *practiced* and achieved expert-level performance within the narrow domain of playing *Jeopardy!* Watson represents a new kind of computer system called a cognitive system (Wladawsky-Berger, 2013; Isaacson, 2014; Feldman and Reynolds, 2014; Feldman, 2016).

3.3 The Human/Cog Ensemble

The primary goal of most of artificial intelligence research has been to replicate human intelligence with the idea to ultimately compete with or replace humans. Indeed, much has been written recently about the fear that AI will take over and make humans obsolete. The cognitive augmentation era will be different. Instead of replacing humans, cognitive systems seek to act as partners with and alongside humans. John Kelly, Senior Vice President and Director of Research at IBM describes the coming revolution in cognitive augmentation as follows (Kelly and Hamm, 2013):

> *"The goal isn't to ... replace human thinking with machine thinking. Rather ... humans and machines will collaborate to produce better results—each bringing their own superior skills to the partnership. The machines will be more rational and analytic—and, of course, possess encyclopedic memories and tremendous computational abilities. People will provide judgment, intuition, empathy, a moral compass and human creativity."*

Interestingly, over thirty years ago, Apple, Inc. envisioned an intelligent assistant called the *Knowledge Navigator* (Apple, 1987). The Knowledge Navigator was a vision of an artificial executive assistant capable of natural language understanding, independent knowledge gathering and processing, and high-level reasoning and task execution. Totally fiction, the Knowledge Navigator concept was well ahead of its time and many people derided Apple over the idea. However, we are beginning to see some of the features in current voice-controlled "digital assistants" such as Siri, Cortana, and Amazon Echo and now we are also beginning to see cognitive systems beginning to be able to perform high-level cognition.

In 2014, IBM released a video demonstrating humans collaborating with an advanced version of IBM Watson technology (Gil, 2014). Some aspects of the video are strikingly similar to the Knowledge Navigator video of 1987, particularly the collaborative nature of the dialog. However, the Watson technology shown in the video was real. In the video, Watson communicates with the humans in natural spoken language. Watson also performs various manipulations of data and information presenting the results to the humans. At one point one of the humans ask Watson to form a conclusion and make a recommendation which it does.

The collaborative nature of one or more humans working with artificial entities is key to the coming cognitive systems era as depicted in Fig. 3-2. In the spirit of Engelbart's framework, we draw the human and the artificial entity ("cog") as two components working together as a single system— an ensemble. The dashed line represents the border between the human/cog ensemble and the outside world. Data, information, knowledge, and wisdom flow from the outside into the ensemble, and transformed data, information, knowledge, and wisdom flow out of the ensemble.

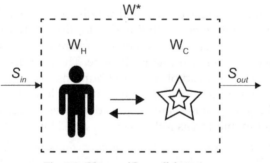

Fig. 3-2: Human/Cog collaboration.

3.4 The Cognitive Process

Viewing the human/cog ensemble as a transformer of information is an important concept. All cognitive systems, whether they are human or artificial, process and transform information. Information becomes more valuable as it is processed and combined with other pieces of information and we assign these forms of information different names: *data, information, knowledge,* and *wisdom*. The DIKW hierarchy, shown in Fig. 3-3, is a well-respected idea from the knowledge management field representing

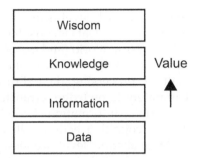

Fig. 3-3: The DIKW hierarchy.

information as processed data, knowledge as processed information, and wisdom as processed knowledge (Ackoff, 1989).

A similar depiction of the knowledge hierarchy is seen in the National Security Agency (NSA) Reference Model shown in Fig. 3-4 (Hancock et al., 2019).

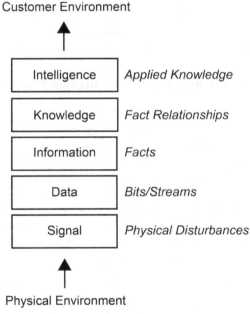

Fig. 3-4: The RSA reference model.

Because this depiction comes from the intelligence community, the NSA Reference model calls "Wisdom" "Intelligence" and also depicts the *signal* level. Physical disturbances in the environment create detectable signals. Once perceived, these signals can be turned into data feeding the hierarchy.

In the knowledge hierarchy, each level is of a higher value than the lower level because of the processing. Data is considered to be of the lowest value and the closest to the physical world (and therefore the least abstract of the levels). Data is generated when physical phenomena are sensed. Information is the result of processing data. Processing information produces knowledge. Ultimately, knowledge is transformed into wisdom.

The transformation of data, information, knowledge, and wisdom is the essential aspect of a cognitive action called a *cognitive process* (Fulbright, 2017b; 2018). At an abstract level, we view a cognitive process as receiving data, information, knowledge, or wisdom as an input and producing transformed data, information, knowledge, or wisdom as an output as show in Fig. 3-5.

The execution of every cognitive process requires the expenditure of a quantity of *cognitive work, W,* to transform the input, S_{in} into the output, S_{out}.

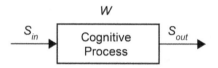

Fig. 3-5: A cognitive process as a transformation of information.

3.5 Cognitive Work and the Augmentation Factor

A human/cog ensemble, shown in Fig. 3-2, is equivalent to the cognitive process shown in Fig. 3-5. The human/cog ensemble is an information-processing system performing an overall cognitive process by executing several internal cognitive processes. In a human/cog ensemble, the human performs some of the cognitive processing and expends W_H amount of cognitive work. The cog performs some of the cognitive processing expending W_C. Together, the ensemble expends a total amount of cognitive work, W^*. The total cognitive work performed by the ensemble is at least equal to the sum of the cognitive work done by each component

$$W^* = W_H + W_C \qquad (3.1)$$

There is speculation, and reason to believe, W^* could actually exceed the sum due to emergent properties arising out of the human/cog interaction. However, this has not be proven and is the subject of future work.

Given we can calculate the individual cognitive contributions of the human and the cog, comparing the efforts yields a metric called the *augmentation factor, A^+*:

$$A^+ = \frac{W_C}{W_H} \qquad (3.2)$$

Humans working alone without the aid of artificial entities are not augmented at all and have an $A^+ = 0/W_H = 0$. If humans are performing more cognitive work than artificial entities, $A^+ < 1$. This is the world in which we have been living so far. However, when cogs start performing more cognitive work than humans, $A^+ > 1$ with no upward bound. That is the coming cognitive systems era.

3.6 Cognitive Power

In physics and engineering, it is a common practice to measure the amount of work performed over the amount of time it takes to perform the work. Therefore, *cognitive power*, P, is the execution of cognitive work over a period of time,

$$P = \frac{W}{t}. \tag{3.3}$$

The cognitive power of the human/cog ensemble is the combination of the cognitive power contributions of the human and the cog,

$$P^* = \frac{W^*}{t}. \tag{3.4}$$

The goal of human cognitive augmentation is for the human to be able to perform a larger amount of cognitive work, and thereby exhibit a greater amount of cognitive power, by collaborating with one or more cogs as shown in Fig. 3-6.

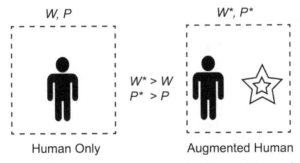

Fig. 3-6: Human cognitive augmentation measured with cognitive work and cognitive power.

3.7 Cognitive Accuracy and Cognitive Precision

Another way to measure the effect of cognitive processing done by the human/cog ensemble is to measure the effect on *cognitive accuracy* and *cognitive precision* (Fulbright, 2019).

To define the notions of *cognitive accuracy* and *cognitive precision*, we first model the human/cog ensemble as a general information machine

(GIM) (Fulbright, 2002). Formally, a GIM is a stochastic Turing machine accepting information as an input and producing information as an output. In a traditional Turing machine, rules specify symbol transitions dictating a deterministic transformation of an input to a specific output. However, the rules in a GIM are stochastic in nature. Each transition rule is associated with a probability rather than being a deterministic certainty. Therefore, for a given input, the GIM's output may vary with each run. Over a number of runs, given the same input, a set of outputs (C) is created. The pattern of outputs in C is determined by the randomness of the probabilities within the GIM and denoted as λ as shown in Fig. 3-7.

If the GIM is truly random, the outputs in C are evenly distributed with the average probability of each output, c, being $1/|C|$ where $|C|$ is the cardinality of C or simply the number of different outputs. If the GIM is truly deterministic (such as that of a deterministic Turing machine), one and only one output will be generated 100% of the time. Of course, the probability of that output is 100% and the probability of any other possible output in C is zero.

If, however, the randomness of the GIM is an intermediate value a pattern of outputs will emerge. Some of these outputs will be very similar to other outputs and can be grouped together into subsets of C. The

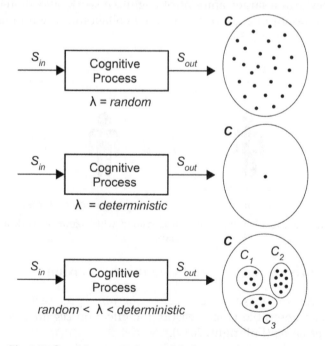

Fig. 3-7: Cognitive processing modeled as a stochastic process.

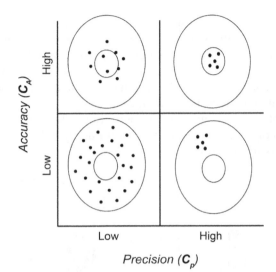

Fig. 3-8: Precision and accuracy.

probability of a subset can be calculated by comparing the cardinality of the subset with the cardinality of *C* as shown in Fig. 3-7.

The distribution pattern of *C* is critical. If we choose one of the subsets in *C* as the *preferred* or *desired* output, we can characterize any distribution pattern based on the ideas of *accuracy* and *precision* as shown in Fig. 3-8. *Cognitive accuracy* involves the propensity to produce the preferred output. *Cognitive precision* involves the propensity to produce *only* the preferred output.

The goal, of course, is for every output to fall within the preferred subset (upper right quadrant). This represents high accuracy and high precision. It is possible for outputs to be very similar to each other (forming a tight cluster) but not falling within the preferred subset (lower right quadrant). This represents high precision but low accuracy. Outputs centered on the preferred subset but not tightly clustered (upper left quadrant) represents high accuracy but low precision. Outputs with low accuracy and low precision (lower left quadrant) have only accidental relationship to the preferred subset.

Since the outputs in the model are the result of cognitive processing, we call these two measures *cognitive accuracy* (C_A) and *cognitive precision* (C_P). The result of any cognitive process can be either the desired result (or close to it) or an undesired result. We define *cognitive accuracy* (C_A) as the propensity to produce the desired result. We define *cognitive precision* (C_P) as the propensity to not produce something other than the desired result.

Note, these are not necessarily equivalent to "correct" and "incorrect" results. Often, the result of cognitive processing cannot be labeled as correct or incorrect. For example, asking a person what things in life are important to them will generate a number of answers. It is not possible to determine if one of those answers is correct and the rest incorrect. However, we can identify a particular answer as being the one we desire. Once we have chosen the target, we can calculate accuracy and precision of any set of answers relative to the target.

The goal of human cognitive augmentation is to increase the cognitive accuracy and cognitive precision of the human by collaborating with one or more cogs as shown in Fig. 3-9.

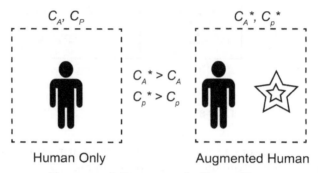

Fig. 3-9: Human cognitive augmentation measured with cognitive accuracy and cognitive precision.

3.8 Levels of Cognitive Augmentation

We think it will be many years before fully artificial intelligences become available to the mass market. In the meantime, there will be human/cog ensembles with varying amounts of cognitive augmentation. In some ways, the evolution of cogs is similar to the evolution of self-driving automobile technology. The National Highway Traffic Safety Administration (NHTSA) has defined six levels of automation (NHTSA, 2019). These levels go from Level 0 (no automation) to Level 5 (full automation). Intermediate levels include Level 1, where the driver is assisted by a single automated system, Level 2, where the vehicle controls some functions but the human can intervene at any time, and Level 3, where the vehicle makes decisions and performs complex functions usually without human intervention but the human can intervene if needed.

Taking this as inspiration, we define the following Levels of Cognitive Augmentation ranging from no augmentation at all (all human thinking) to fully artificial intelligence (no human thinking):

Up until this point in time, computers and software we have used represent Level 1 cognitive augmentation—assistive tools. Recent advances in deep learning and unsupervised learning have produced

Level 0: No Augmentation
  human performs all cogntiive processing

Level 1: Assistive Tools
  abacus, calculators, software, etc.

Level 2: Low-Level Cognition
  pattern recognition, classification, speech
  human makes all high-level decisions

Level 3: High-Level Cognition
  concept understanding, critique,
  conversational natural language

Level 4: Creative Autonomy
  human-inspired, unsupervised synthesis

Level 5: Artificial Intelligence
  no human cognitive processing

Fig. 3-10: Levels of cognitive augmentation.

Level 2 cognitive augmentation. But as the abilities of cogs improves, we will see Level 3 and Level 4 cognitive augmentation. Later, we introduce the Model of Expertise and show how it can produce Level 3 and Level 4 cogs.

3.9 Summary

Technology has always changed humankind. As new technology is adopted, social, cultural, political, and ethical norms change. Technology augments human capability. Throughout history, most new technology has augmented humans in a physical sense. Information technology augments human cognitive ability. Until now, cognitive augmentation technology has only made humans better thinkers. Humans have still done all of the thinking. However, cognitive systems technology is producing artificial systems capable of performing some of the thinking on their own.

In the coming cognitive systems era, humans will collaborate with these systems. There will be at least two entities doing the thinking (humans will partner with more than one cognitive system). Therefore, cognition will be the product of a collaborative effort. The goal is for humans to be able to perform higher levels and higher amounts of cognition by virtue of this collaboration. We can measure the effect of cognitive augmentation by the amount of cognitive work performed, the amount of cognitive power expended, and increases in the cognitive accuracy and cognitive precision achieved.

The level of performance of the human/cog ensemble will be indistinguishable from, and even exceed, the level of an expert. At that point, we will have achieved *synthetic expertise.* When synthetic expertise is adopted by the majority cultural and societal changes will follow.

Chapter **4**

A Brief History
of Artificial Intelligence

The field of artificial intelligence has enjoyed a long history dating back more than 65 years but extending several decades before.

4.1 The Dartmouth Conference

The field of artificial intelligence (AI) is largely regarded to have started as a result of a summer project at Dartmouth in 1956. In 1955, John McCarthy proposed the seminar to investigate how a machine can simulate intelligence (McCarthy et al., 1955):

> We propose that a 2-month, 10-man study of artificial intelligence be carried out during the summer of 1956 at Dartmouth College in Hanover, New Hampshire. The study is to proceed on the basis of the conjecture that every aspect of learning or any other feature of intelligence can in principle be so precisely described that a machine can be made to simulate it.

Among the attendees were many pioneers and luminaries of AI:

- John McCarthy (developer of the LISP programming language)
- Marvin Minsky (the first randomly wired neural network, framework for knowledge representation, the society of mind theory, and the emotion machine theory of intelligence)
- Allen Newell (developer of the first AI program, the Logic Theorist, the Soar architecture, and the Knowledge Level)
- Herbert Simon (the Logic Theorist and General Problem Solver, theory of human problem solving)

- Arthur Samuel (seminal work on machine learning, checkers, game status scoring)
- Claude Shannon (information theory)
- Ray Solomonoff (algorithmic information theory, inductive inference)
- Oliver Selfridge (machine perception)
- Ross Ashby (complex systems, intelligence amplification)

Also in attendance were several engineers who would go on to build the first generation of computers as we know them today.

The Dartmouth conference was an attempt to organize and clarify the many ideas about machines and intelligence existing at the time. Several attendees of the Dartmouth conference were from the field of cybernetics—a field of study dating back well into the early 1800s. Weiner (1948) defines cybernetics as the study of control in the animal and the machine. Among the topics cyberneticists sought to understand were *cognition* and *learning*.

In defining computability, Turing (1936) described what would become known as Turing machines. A Turing machine is an abstract machine, able to manipulate information in the form of symbols, and simulate any algorithm. Shannon (1948) established the field of information theory—the study of the quantification, storage, and communication of information—but Hartley (1928) and Nyquist (1924; 1928) introduced the idea of information as a quantifiable and measurable quantity. Bush (1945) imagined extending and enhancing human memory with an artificial associative memory system called the Memex. Turing (1950) discussed if machines themselves could think and offered the "imitation game" (later to be known as the "Turing Test") to decide if a machine is exhibiting intelligence. McCarthy et al. (1955) coined the term *artificial intelligence*.

Topics explored at the Dartmouth conference included: *computers, natural language processing, neural networks, theory of computation, abstraction, and creativity*. The initial outlook in the mid-1950s was optimistic with Herbert Simon declaring "machines will be capable, within twenty years, of doing any work a man can do" and Marvin Minksy predicting "within a generation ... the problem of creating 'artificial intelligence' will substantially be solved" (Gaskin, 2008).

Newell and Simon formulated the notion of symbol manipulation early in AI research and the idea persists to present day. The Physical Symbol System hypothesis (PSS) maintains knowledge can be represented by symbols which can be combined into structured expressions (predicate statements, production rules) and manipulated (Newell and Simon, 1976). Newell and Simon believed symbol manipulation was necessary for intelligence. Most AI systems and cognitive architectures have some form of symbolic representation and manipulation.

Another notion from early in AI research is the neural network model (McCulloch and Pitts, 1943; Hebb, 1949; Medler, 1998). Contrary to representing knowledge as discrete symbols, knowledge is represented in neural networks as the network's response to stimuli. The "connectionist" model and the "symbol system" model have persisted throughout the history of AI, sometimes clashing and sometimes complementing each other. The human mind features both paradigms. The human brain is a neural network. However, when thinking, the human brain manipulates symbols. The idea of understanding how the human mind works and mimicking it in an artificial system has driven most of AI research and development over its 65-year history.

In this chapter we present a brief history of artificial intelligence. We readily acknowledge not every contribution to the field is represented here. Later in this book, we introduce novel Model of Expertise and describe in some detail several applications of the model. The systems we describe heavily rely on several key technologies from artificial intelligence: machine learning, deep learning, and unsupervised learning, natural language processing and conversational interfaces, and high-level reasoning. The contributions throughout the history of artificial intelligence mentioned in this chapter focus on these key technologies.

4.2 Milestones in Artificial Intelligence

The Logic Theorist, written in 1956 by Allen Newell, J.C. Shaw, and Herbert Simon is generally regarded as the first AI program (Newell and Shaw, 1956). The purpose of Logic Theorist was to show a machine could prove theorems as well as a human mathematician. The program did this by representing different logical deductions as branches of a tree. The task then became a tree search process. Heuristics, rules of thumb, were used to build the tree efficiently and list processing was used to implement the search and processing in the programming language. Tree search, heuristics, and list processing became standard practices in AI and are still important today.

Based on list processing, John McCarthy created the LISP programming language in 1958 (McCarthy, 1960). LISP has been a staple of artificial intelligence research ever since. Many dialects of LISP have been created over the decades. LISP and its central data structure, linked lists, were ideal for representing and processing symbols and knowledge represented by symbols (Newell and Simon's PSS).

Ross Ashby coined the term *intelligence amplification* in 1956 maintaining human intelligence could be synthetically enhanced by increasing the human's ability to make appropriate selections on a persistent basis (Ashby, 1956). But again, the human does all of the thinking. The synthetic aids just make the human more efficient. That computers could augment human activity seems trivial to us today, but this kind of thinking was new at the beginning of the artificial intelligence era and lead to Licklider and Engelbart's human/computer symbiosis ideas in the 1960s (Licklider, 1960; Engelbart, 1962).

Following work started with the Logic Theorist, the General Problem Solver (GPS) was demonstrated by Newell, Shaw, and Simon from 1957–1959 and is a prime example of symbol manipulation (Newell et al., 1959). Any problem able to be expressed as a set of symbols from a language, defining the symbols and operation on those symbols, can be solved, in theory, by GPS. Practically, however, GPS did not perform well on real-world problems because such problems tend to result in exponential explosion of clauses to capture the knowledge. GPS automatically performed the symbol manipulation but had no understanding of the meaning of the symbols. Humans attached semantic meaning to the symbols. However, GPS was the first computer program separating its knowledge of problems from the strategy of how to solve problems and was the precursor to expert systems of the 1970s and 1980s, and the Soar cognitive system architecture.

One of the first systems to achieve a level of expertise, SAINT (Symbolic Automatic INTegrator), was created in 1961 by James Slagle (Slagle, 1961). SAINT solved symbolic integration problems in freshman calculus. Integration in calculus is largely a process of manipulating symbols based on a set of rules, or heuristics, and so lends itself nicely to a symbolic manipulation program like SAINT. Written in LISP, SAINT achieved proficiency at roughly the high-school level.

In 1965, Joseph Weizenbaum built ELIZA, an interactive program able to simulate a dialogue in English language on any topic (Weizenbaum, 1966). ELIZA did not understand conversations but, by using a pattern matching and substitution, the program could recreate the feel of a human-human dialog. As such, ELIZA was one of the first forays into natural language processing (NLP) and was a precursor to today's chatbots. Interestingly, even though ELIZA had no understanding, humans attributed human-like feelings to the program and were convinced it truly demonstrated intelligence and understanding.

In 1965, Edward Feigenbaum, Bruce Buchanan, Joshua Lederberg and Carl Djerassi initiated Dendral, a multi-year effort to develop software to deduce the molecular structure of organic compounds using scientific instrument data (Lindsay et al., 1980). Dendral is considered the first expert system because human experts' knowledge and know-how were captured into a knowledge base consisting of knowledge and heuristics. Programming written in LISP manipulated this knowledge and automated the decision-making process and problem-solving behavior of organic chemists.

Dendral lead to several other programs of note in artificial intelligence. For example, MYCIN was developed ten years later in 1974 by Edward Shortliffe demonstrating a practical rule-based approach to medical diagnoses, even in the presence of uncertainty (Buchanan and Shortliffe, 1980). While it borrowed from DENDRAL, its own contributions strongly influenced the future of expert system development, especially commercial systems. In 1975, the Meta-Dendral learning program produced new results in chemistry resulting in the first scientific discoveries by a computer to be published in a refereed journal.

One of the first successful commercial systems was Macsyma initially developed in 1968 by Carl Engelman, William A. Martin, and Joel Moses (Moses, 2008; 2012). The first successful knowledge-based program in mathematics, Macsyma was able to manipulate algebra expressions in a manner similar to humans. Macsyma became a commercial product in the 1990s and is still in use today.

In 1971, Terry Winograd's PhD thesis demonstrated the ability of computers to understand English sentences in a restricted world of children's blocks, in a coupling of his language understanding program, SHRDLU, with a robot arm carrying out instructions typed in English (Winograd, 1971; 1972).

Herbert A. Simon won the Nobel Prize in Economics in 1978 for his theory of bounded rationality, one of the cornerstones of artificial intelligence known as "satisficing" (Simon, 1956). Satisficing describes a decision-making strategy to find satisfactory solutions (not necessarily optimum solutions) in realistic situations involving partial, missing, or contradictory information.

In 1979, the Stanford Cart, built by Hans Moravec, became the first computer-controlled, autonomous vehicle when it successfully traversed a chair-filled room and circumnavigates the Stanford AI Lab (Moravec, 1990).

John Laird and Paul Rosenbloom, working with Allen Newell, completed CMU dissertations on Soar in 1983. The Soar architecture would become the most successful cognitive architecture and is still in development today (Laird, 2012).

During the 1980s, expert systems using if-then production rules sought to capture experts' knowledge and simulate the logical inference of experts in a field of discipline (Hayes-Roth et al., 1983). These systems required laborious programming and knowledge engineering. Expert systems proved to be brittle in the face of uncertain or conflicting information.

DARPA has been a primary funding source for artificial intelligence for several decades. Introduced in 1991, DART (Dynamic Analysis and Replanning Tool), the scheduling application deployed in the first Gulf War, paid back DARPA's 30-year investment in artificial intelligence research by 1995. DART uses intelligent agents and artificially intelligent planning and optimization algorithms (Reece-Hedberg, 2002).

In the mid-1980s, the first robotic cars were created by Ernst Dickmanns and his group at the University Bundeswehr Munich (UniBW) in Munich, Germany using machine vision. By the late 1980s and early 1990s, more than one billion dollars were spent on the pan-European Prometheus project (Schmidhuber, 2019). In 1994, with passengers on board, the twin robot cars built by Dickmanns' group, VaMP and VITA-2, drove more than one thousand kilometers on a Paris three-lane highway in standard heavy traffic at speeds up to 130 km/h. They demonstrated autonomous driving in free lanes, convoy driving, and lane changes left and right with autonomous passing of other cars (Dickmanns, 2007).

In the late 1980s through the mid 1990s, Dean Pomerleau and Todd Jochem at Carnegie Mellon created ALVINN (An Autonomous Land Vehicle in a Neural Network). ALVINN's neural network-based vision-controlled steering has become the basis for today's self-driving cars (Jochem et al., 1995).

In 2002, iRobot' launched the Roomba, able to autonomously vacuum floors while navigating and avoiding obstacles. iRobot was founded in 1990 by MIT roboticists Colin Angle, Helen Greiner and Rodney Brooks. While not artificially intelligent, for many, Roomba is the first exposure to autonomous devices in the home.

In 1997, IBM's Deep Blue defeated the human chess champion Gary Kasparov (IBM, 2018). Chess had long been an objective of artificial intelligence researchers going back to the 1950s. Even though many argued Deep Blue took more of a brute-force approach than an artificial intelligence approach, the victory over the human chess champion gained international interest much like victories in *Jeopardy!* and Go would later.

In 2000, Cynthia Breazeal at MIT published her dissertation on sociable machines, describing Kismet, a robot with a face able to express emotions (Breazeal, 2000). The social nature of cognitive systems is expected to be a major component of the cog era as humans will form emotional connections with cognitive systems as they evolve and take on personalities much like pets.

In 2004 and 2005, the Defense Advanced Research Projects Administration (DARPA) held the DARPA Grand Challenge, a prize competition for autonomous vehicles (DARPA, 2019). No vehicles completed the 150-mile course in 2004 but five vehicles completed the 2005 competition with the Stanford University team taking first place and the Carnegie Mellon team taking second place. In 2007, DARPA held the DARPA Urban Challenge requiring autonomous vehicles to navigate a 60-mile course in an urban setting, obeying traffic laws, merging with other traffic, detecting and avoiding other obstacles. The Stanford University team won first place and the Carnegie Mellon team took second place. Google introduced its own self-driving car in 2009. Many car makers are now developing their own self-driving technology.

In 2011, Apple released its Siri technology for the iPhone (Apple, 2015). Siri is a virtual assistant able to understand natural-language voice commands and reply in spoken natural language. While at one time, speech synthesis and speech recognition were sought-after targets of artificial intelligent research, today no one considers speech recognition an example of artificial intelligence. Since Siri's release in 2011, most major handheld device manufacturers have released their own version of voice-activated digital assistants: Google's Google Now (Google, 2015), Microsoft's Cortana (Microsoft, 2015), Amazon's Echo (Alexa) (Colon and Greenwald, 2015). To date, billions of people have experienced cognitive systems and artificially intelligent technology through text-based chatbots and voice-activated virtual assistants. This is likely to continue to be a major touch point for cognitive systems technology as it evolves higher-level cognitive abilities.

In 2011, IBM's Watson defeated two of the most successful human champions in the game of *Jeopardy!* (Wladawsky-Berger, 2013; Jackson, 2011). For many, Watson's victory over two of the best human champions of all time marks the beginning of the "cognitive system era." Watson used a number of techniques from the artificial intelligence and machine learning fields. Watson launches multiple searches for information in parallel and then aggregates and scores results until a consensus answer emerges and exceeds a certain confidence factor (Ferrucci et al., 2010; Ferrucci, 2012). The process closely mimics the way the human mind reasons, so represents an artificial system performing the equivalent of higher-order cognition. For this reason, IBM and others call Watson-like technology *cognitive systems* (Kelly and Hamm, 2013).

In 2012, a convolutional neural network called AlexNet achieved an error of 15.3% in the annual ImageNet contest beating the next-best competitor by over 10 percentage points (Krizhevsky et al., June 2017). This victory got the attention of the machine learning and artificial intelligence communities but also garnered significant interest from others outside of the artificial intelligence field and is seen as the catalyst of the recent resurgence of artificial intelligence. Deep learning became a household world in artificial intelligence (Economist, 2016).

In much the same way DARPA did for autonomous vehicles earlier, from 2012 to 2015, DARPA held the DARPA Robotics Challenge, a prize competition focusing on emergency-response scenarios (DARPA, 2019b). Robotic systems had to drive to the situation, travel across rubble, remove debris, climb ladders, use tools to break through concrete, locate and close a leaky valve, and connect a firehose. The South Korean Team KAIST tied for first with the Florida Institute for Human and Machine Cognition (IHMC) and the Carnegie Mellon/General Motors Team Tartan.

In 2013, NEIL, the Never Ending Image Learner, was released at Carnegie Mellon University to constantly compare and analyze relationships between different images (Chen et al., 2013). NEIL uses semi-supervised machine learning to discover relationships in unstructured data such as images found on the Internet. Semi-supervised and unsupervised learning, as described later, are important areas for artificial intelligence and cognitive systems because it permits the system to learn on its own at computer speeds. The cognitive systems era is likely to be ruled by self-taught systems.

As a case in point, in 2016, Google's AlphaGo, developed by DeepMind Technologies (subsequently purchased by Google), defeated the reigning world champion in Go (DeepMind, 2018a). In 2017, an even stronger version called AlphaGo Master won 60 online games against professional human players over a one-week period. Also in 2017, a version called AlphaGo Zero learned how to play Go by playing games with itself and not relying on any data from human games (DeepMind, 2018b). A generalized version called AlphaZero was developed in 2017 capable of learning any game. While Watson required many person-years of engineering effort to program and teach the craft of Jeopardy, AlphaZero achieved expert-level performance in the games of Chess, Go, and Shogi after only a few hours of unsupervised self-training.

4.3 Machine Learning

Enabling a machine to learn has been a goal of AI researchers since the beginning of the AI field (Bishop, 2006). Dartmouth conference attendee, Arthur Samuel, coined the phrase in 1959 (Samuel, 1959). However, Alan Turing, as early as 1947 spoke about machines able to learn from experience (Turing, 1947).

In general, machine learning algorithms must be trained. For most of AI's history, training has been *supervised* meaning the training set was carefully engineered by humans to contain both positive and negative labeled examples of the thing the humans wanted the machine to learn. Engineering of the training set requires significant, and sometimes overwhelming, amount of time and effort. In *semi-supervised learning* labeled training data is used but unlabeled data is also used. This reduces the amount of human engineering required and also opens machine learning up to unstructured data such as text messages, images, sounds, videos, etc. *Unsupervised learning* uses unlabeled data, so requires no human engineering. Unsupervised machine learning is free to identify any patterns in the data sometimes leading to unexpected discoveries.

In 1951, even before the Dartmouth conference, Marvin Minsky, then a graduate student, developed the first neural network machine able to learn, called SNARC (Stochastic Neural Analog Reinforcement Calculator). SNARC was a randomly connected network of approximately 40 synapses each having a memory holding the probability that signal comes in one input and another signal will come out of the output (Crevier, 1993).

In 1952, Arthur Samuel created a computer program to play checkers. Over 40 years later, a program called Chinook would defeat the world champion in checkers (Samuel, 1959).

An analog device called MENACE (Machine Educable Noughts and Crosses Engine) was developed in 1960 by Donald Michie able to learn to play a perfect game of tic-tac-toe. Instead of electronic circuits, MENACE used matchboxes as nodes and beads as weights (Michie, 1963).

The perceptron, invented by Frank Rosenblatt in 1958, was able to learn a set of weights, attached to each node in a highly-connected network, needed to classify inputs (images originally) into positive and negative classes (Rosenblatt, 1958). Initially, perceptrons used only one layer, limiting the ability of the network to learn. Later, multi-layer convolutional neural networks would overcome limitations especially when backpropagation techniques were used. Limitations of early efforts in machine learning and lack of results lead to most researchers abandoning of machine learning through much of the 1970s and into the 1980s.

In 1970, Patrick Winston's program, ARCH, learned concepts from examples in the world of children's blocks (Winston, 1970). ARCH represented relationships between structural elements in a semantic network and represents an early form of concept learning. Multi-level concept learning in deep-learning networks would become important forty years later.

In 1970, Seppo Linnainmaa published the general method for automatic differentiation (Linnainmaa, 1976). This corresponds to the modern version of backpropagation to come later—a family of methods used to train artificial neural networks by efficiently calculating and updating the weights. Backpropagation and reinforcement learning later would become essential to high-profile machine learning achievements.

Having originally suggesting genetic algorithms in 1960, John Holland, one of the pioneers of artificial intelligence and complex adaptive systems, published a book on genetic algorithms in 1975 called *Adaptation in natural and artificial systems* (Holland, 1975). Genetic algorithms rely on a large number of successive generations of evolving systems, somewhat like training iterations of neural network-based learning systems. Like in Darwin's theory of evolution, the performance of each instance in a generation is evaluated and only the strongest allowed to continue on to the next generation. Each generation is slightly different than the previous generation and different instances within a generation employ different mechanisms as well. A genetic algorithm system can explore many different possible paths and "learn" or "evolve" a superior system.

In 1979, Kunihiko Fukushima published work on the neocognitron, a type of multi-layered artificial neural network (ANN) capable of visual pattern recognition. The neocognitron later inspired convolutional neural networks (CNN)—a major step forward in machine learning when combined with backpropagation (Fukushima, 1980).

In 1981, Gerald Dejong introduced Explanation Based Learning (EBL)—a computer algorithm able to discard irrelevant data and create generalized rules (DeJong, 1981; DeJong and Lim, 2017). EBL used domain knowledge to improve supervised learning. The domain knowledge represented an expert's imperfect and incomplete understanding of complex world behavior relative to the domain. Inference over the domain knowledge generates additional domain knowledge. By this mechanism, a cognitive system may learn new things about a domain of discourse on its own. Domain knowledge, described later in this book, is an important feature of the Model of Expertise (Fulbright, 2020).

In 1982, John Hopfield popularized Hopfield networks. Hopfield networks connect nodes along a directed graph in a temporal sequence enabling the network to maintain information over a time sequence—a form of content-addressable associative memory inspired by the human brain (Hopfield, 1982). In 1986, David Rumelhart, Geoffrey Hinton, and Ronald Williams applied Linnainmaa's back propagation to multi-level neural networks and showed the ability to learn data representations (Rumelhart et al., 1986). This would lead to a type of neural network called recurrent neural networks. Recurrent neural networks are good for analyzing time-based inputs such as speech and handwriting where data at a specific point in time depends on what comes immediately before or immediately after.

In 1985, Terrence Sejnowski and Charles Rosenberg invented NetTalk, an artificial neural network able to construct simplified models (Sejnowski and Rosenberg, 1987). NetTalk learned associations between correct pronunciations and the context in which the letters appear and therefore is somewhat similar to time-sequence behavior in recurrent neural networks.

In 1986, Rina Dechter introduced the term *deep learning* (Dechter, 1986). Deep learning would cause a revolution in machine learning and a resurgence of artificial intelligence some twenty years later. Deep learning uses multiple layers of neural networks to progressively extract higher level features from the input data. This heralded a new age of data analytics in the 1990s and 2000s.

In 1989, Q-learning was developed by Christopher Watkins and represents a practical form of reinforcement learning (Watkins, 1989). In considering a sequence of actions leading from a starting state to an end state, Q-learning seeks to maximize the expected value of total reward over any successive steps where the "reward" is a measure of the quality of the action. The algorithm seeks to learn the most profitable sequence of steps. Unlike supervised learning, Q-learning does not require a carefully engineered and labeled training set. Reinforcement learning figures prominently into recent achievements in deep learning, particularly in the DeepMind systems such as AlphaGo, AlphaGo Zero, and AlphaZero.

In the 1990s, researchers began using a data-driven approach for machine learning. Instead of basing learning on carefully engineered training knowledge, researchers starting creating systems to analyze large amounts of semi-structured and even unstructured data. The goal was to draw conclusions based on the data an analysis by detecting associations and patterns present in the data. Later, this would become the field of *data analytics* and would also help usher in the *deep learning* era.

In 1992, Gerald Tesauro developed TD-Gammon using an artificial neural network trained using temporal-difference learning (Tesauro, 1995). TD-learning is similar to Q-learning in that seeks to calculate a total expected reward value over successive steps. TD-Gammon was able to rival top human backgammon players but did not defeat the human champion.

In 1995, Kam Ho published a paper describing random decision forests (Ho, 1995). A decision tree is a tree-like data structure, constructed during supervised training iterations, representing decisions and possible consequences. Decision trees have been instrumental in data mining and machine learning where pattern detection and trend detection is important.

In 1995, Corinna Cortes and Vladimir Vapnik published work regarding support vector machines (SVM). Given a labeled training set, SVM is an algorithm able to calculate the best dividing line between two sets of data, even with multiple features (Cortes and Vapnik, 1995). SVMs have been critical in data mining, data analytics, and machine learning systems.

In 1997, IBM's Deep Blue defeated the reigning human world chess champion Kasparov (IBM, 2018). Deep Blue was built by IBM especially for playing chess and included special hardware and software just for playing chess. However, IBM also learned by playing hundreds of games with human trainers. At the time, this was considered to be a major achievement in the field of artificial intelligence.

In 1997, Sepp Hochreiter and Jürgen Schmidhuber developed long short-term memory (LSTM), a form of recurrent neural network (Hochreiter and Schmidhuber, 1997). LSTMs are able to store data of various types at nodes and process the data to calculate the weights associated with one or more connections to other nodes with the weights being learned over multiple training iterations. LSTM greatly improved the efficiency and practicality of recurrent neural networks which have been used in speech recognition, machine translation, and handwriting recognition.

In 2006, the Netflix Prize competition was launched challenging researchers to use machine learning to beat the accuracy of Netflix's own recommendation software. The prize was won in 2009 by Bellikor's Pragmatic Chaos (Buskirk, 2009). Bellikor was a hybrid team formed by AT&T research and Australian team Big Chaos.

In 2009, ImageNet was created by Fei-Fei Li of Stanford University. The ImageNet database contains more than 14 million hand-annotated images indicating what objects are pictured and containing over 20,000 categories. Since 2010, a competition known as the ImageNet Large Scale Visual Recognition Challenge (ILSVRC), has been held challenging systems to correctly classify and detect objects and scenes. The contest has been seen as a barometer measuring the progress of deep learning. Early error rates were in the vicinity of 25% but by 2017 error rates of 5% or lower were regularly obtained, rivaling or exceeding human performance (Krizhevsky et al., June 2017). In 2015, Microsoft's convolutional neural network with over 100 layers won the ImageNet 2015 contest with an error rate of only 3.57% (Linn, 2015).

In 2011, IBM's Watson defeated two of the best human champions on the television quiz show Jeopardy! (Markoff, 2011; Jackson, 2011; Wladawsky-Berger, 2013). Watson was designed to do deep analysis in a generic sense so as to be applicable to problems other than playing games and was programmed to learn how to play Jeopardy! and did so in hundreds of training games (Ferrucci et al., 2010; Ferrucci, 2012). Since 2011, IBM has commercialized Watson-based technologies in a number of different industries and applications (Kelly and Hamm, 2013; Rometty, 2016; Gugliocciello and Doda, 2016; Haswell and Pelkey, 2016; IBM, 2014; IBM, 2015a; IBM, 2015b; Jancer, 2016; Moscaritolo, 2017; Takahashi, 2015; Upbin, 2013).

In 2011/2012, The Google Brain team, led by Andrew Ng and Jeff Dean, created a neural network able to learn to recognize cats by watching unlabeled images taken from frames of YouTube videos (Le et al., 2012). This achievement received broad interest and coverage in the popular press. This is one of the earliest of unsupervised learning using unlabeled training data—a prominent feature of more recent achievements such as DeepMind's AlphaZero (Chessbase, 2018; DeepMind, 2018a; DeepMind, 2018b).

In 2012, a convolutional neural network called AlexNet achieved an error of 15.3% in the ImageNet competition beating the next-best competitor by 10.8 percentage points. AlexNet was developed by Alex

Krizhevsky who was Geoffrey Hinton's, the co-author of the 1986 paper on backpropagation described earlier (Krizhevsky et al., June 2017). This victory got the attention of all in the machine learning and artificial intelligence communities but also garnered significant interest from others outside of the artificial intelligence field and is seen as the catalyst of the recent resurgence of artificial intelligence (Economist, 2016).

In 2014, researchers at Facebook described a neural network called DeepFace able to identify faces with an accuracy over 97% (the FBI's Next Generation Identification System achieves only 85%). DeepFace identifies human faces from digital images using a nine-layer neural network trained on over four million images (Simonite, 2014).

In 2014, Google published information about Sibyl, a machine learning platform to make predictions about user behavior (Woodie, 2014). Google initially developed Sibyl to generate video recommendations for YouTube users. Soon, more people were choosing what video to watch next based on the recommendations than other ways of picking videos. Since then, Google has used Sibyl in a number of other applications such as improving the quality of ads pushed to users and detecting spam and unwanted content in emails. Sibyl uses hundreds of billions of user interactions as its training set.

In 2016, Google's AlphaGo, developed by DeepMind Technologies (subsequently purchased by Google), defeated the reigning world champion in Go (Silver et al., 2016; DeepMind, 2018a). In 2017, an even stronger version called AlphaGo Master won 60 online games against professional human players over a one-week period.

In 2017, a version called AlphaGo Zero learned how to play Go by playing games with itself and not relying on any data from human games (DeepMind, 2018b; Singh et al., 2017). AlphaGo Zero exceeded the capabilities of AlphaGo in only three days and surpassing AlphaGo Master in 21 days.

Also in 2017, a generalized version called AlphaZero was developed capable of learning any game. While Watson required many person-years of engineering effort to program and teach the craft of Jeopardy, AlphaZero achieved expert-level performance in the games of Chess, Go, and Shogi after only a few hours of unsupervised self-training (ChessBase, 2018).

4.4 Game Playing in Artificial Intelligence

Playing games has been an active area for artificial intelligence since its beginning. The well-defined game rules and easy to score performance of game play lends itself nicely to comparing artificial performance to human performance. In 1952, a program called OXO, developed by Alexander Douglas for the EDSAC (Electronic Delay Storage Automatic Calculator) computer, played perfect games of tic-tac-toe against a human opponent (Donovan, 2010). Over the years, computer programs have achieved a level of performance rivaling or surpassing human experts in a number of games.

In 1979, a backgammon-playing computer program developed by Hans Berliner called BKG defeated the reigning world champion (Berliner, 1980). BKG was the first computer program to defeat a human world champion in any board game (Berliner, 1977). Backgammon is different from many board games like checkers and chess because the roll of the dice injects a random factor at each move. In 1992, Gerald Tesauro developed TD-Gammon using an artificial neural network trained using temporal-difference learning (hence the 'TD' in the name) also known as Q-Learning. TD-Gammon was able to rival, but not consistently surpass, the abilities of top human backgammon players (Tesauro, 1995).

In 1995, the computer program Chinook, developed by Jonathan Schaeffer, Rob Lake, Paul Lu, Martin Bryant, and Norman Treloar, defended its Man-Machine World Title against the best checkers player alive at the time. A year earlier, Chinook won the title when Marion Tinsley, the best human checkers player, withdrew from the match due to health reasons. Chinook won the USA National Tournament in 1994 and 1995.

In 1997, Logistello, developed by Michael Buro, defeated the world Othello champion Murakami with a score of 6–0 (NECI, 1997).

Also in 1997, IBM's Deep Blue defeated world chess champion Kasparov. Kasparov had beaten Deep Blue the year before but lost game six in 1997 giving Deep Blue the win (IBM, 2018). The victory was a major milestone in artificial intelligence because chess had long been considered a "holy grail" because playing it at world-class level was considered to require a high degree of intelligence. Deep Blue was built by IBM especially for playing chess taking more than a decade to develop. Deep Blue was the third system in a series of chess-playing systems developed under the direction of Feng-hsiung Hsu. ChipTest was developed at Carnegie Mellon in 1985 followed by DeepThought. DeepThought became the first program to defeat a grandmaster in 1988 but lost to Kasparov in 1989 after which it was renamed and heavily updated over the next eight years leading to the victory over Kasparov. Kasparov has written about the match and humans' interaction with intelligent systems (Kasparov, 2017).

Another milestone was achieved in 2011 when IBM's Watson defeated two of the best human champions on the television quiz show Jeopardy! (Markoff, 2011; Jackson, 2011; Wladawsky-Berger, 2013). Unlike DeepBlue, Watson was designed to do deep analysis in a generic sense so as to be applicable to problems other than playing games. Since 2011, IBM has commercialized Watson-based technologies in a number of different industries and applications (Kelly and Hamm, 2013; Rometty, 2016;

Gugliocciello and Doda, 2016; Haswell and Pelkey, 2016; IBM, 2014; IBM, 2015a; IBM, 2015b; Jancer, 2016; Moscaritolo, 2017; Takahashi, 2015; Upbin, 2013). Watson was programmed to learn how to play Jeopardy! and did so in hundreds of training games. Watson also employed dozens of query, search, and answer aggregation techniques developed in the AI field (Ferrucci et al., 2010; Ferrucci, 2012). Watson could understand written natural language (it can now understand spoken natural language), and answer questions in spoken natural language. It performed millions of actions across thousands of processors in order to formulate the response in time fast enough to beat its human opponents to the buzzer.

In 2016, Google's AlphaGo, developed by DeepMind Technologies, defeated the reigning world champion in Go, Lee Sedol, a game vastly more complex than Chess (Silver et al., 2016; DeepMind, 2018a). Go defies brute-force attempts which rely on calculating and scoring millions of potential moves. Instead, AlphaGo used a Monte Carlo tree search algorithm to find its moves based on knowledge "learned" by machine learning. AlphaGo used an artificial neural network trained extensively both by human and computer play. Thus the tree search gets better which each iteration. An even stronger version called AlphaGo Master won 60 online games against professional human players over a one-week period.

In 2017, a version called AlphaGo Zero learned how to play Go by playing games with itself and not relying on any data from human games (DeepMind, 2018b; Singh et al., 2017). AlphaGo Zero exceeded the capabilities of AlphaGo in only three days and surpassing AlphaGo Master in 21 days. Training AI systems without knowledge engineered from human experts is a significant milestone in AI. Demis Hassabis, the co-founder and CEO of DeepMind, said that AlphaGo Zero was so powerful because it was "no longer constrained by the limits of human knowledge." Removing the need to learn from humans could lead to generalized AI algorithms (Silver et al., 2017).

Also in 2017, a generalized version called AlphaZero was developed capable of learning any game. While Watson required many person-years of engineering effort to program and teach the craft of Jeopardy, AlphaZero achieved expert-level performance in the games of Chess, Go, and Shogi after only a few hours of unsupervised self-training (DeepMind, 2018b).

4.5 Chatbots

A chatbot is an artificial system able to carry on a conversation, usually via textual or auditory means. The ultimate goal of chatbot researchers is to create an artificial system capable of carrying on an extended dialog with a human about any subject so the human would never know he or she was talking to an artificial system. Turing originally proposed the imitation game which over time has become referred to as the Turing Test (Turing, 1950; Saygin et al., 2000). In the most common variation of the Turing Test, a human judge converses in natural language with a human and a machine and must determine which is the human and which is the machine.

The Loebner Prize was established in 1990 by Hugh Loebner and competition has been held annually (AISB, 2019). Using the basic Turing Test formulation, the Loebner prize awards the most human-like systems every year. The $25,000 prize for the system judges cannot distinguish from a human has never been won.

In 2017, Amazon created the annual Alexa Prize to advance human-computer interaction (Alexa Prize, 2019). The goal of the Alexa Prize competition is to create an artificial system able to converse coherently and engagingly with humans on a range of current events and popular topics such as entertainment, sports, politics, technology, and fashion.

One of the earliest successful conversational systems was ELIZA, created from 1964 to 1966 at the MIT Artificial Intelligence Laboratory by Joseph Weizenbaum (Weizenbaum, 1966). Weizenbaum named his program ELIZA after Eliza Doolittle, a working-class character in George Bernard Shaw's Pygmalion taught to speak with an upper-class accent (Weizenbaum, 1976). The program allowed a person to type in plain English at a computer terminal and to interact with a machine in what resembled a normal conversation. Instead of creating and supporting a large, real-world database of information, ELIZA mimicked a Rogerian therapist, frequently reframing a client's statements as questions (Markoff, 2008). ELIZA was featured in a 2012 exhibit at Harvard University titled "Go Ask A.L.I.C.E.," as part of a celebration of mathematician Alan Turing's 100th birthday. The exhibit explored Turing's lifelong fascination with the interaction between humans and computers, pointing to ELIZA as one of the earliest realizations of Turing's ideas (Ireland, 2012).

Weizenbaum contended, despite the fanfare, scripting ELIZA in a way to fool some of the users was relatively easy because Rogerian therapists rely on taking a passive role, and engaging the patient in the conversation by reflecting the patient's statements back by rephrasing them into questions. If nothing else seems to fit the program's scheme, ELIZA always has standard phrases to extend the conversation, such as "Very interesting. Please go on." or "Can you elaborate on that?"

In 1972, Kenneth Colby implemented a chatbot called PARRY (Colby, 1972; Guzeldere, 1995). While ELIZA was a simulation of a Rogerian therapist, PARRY attempted to simulate a person with paranoid schizophrenia PARRY was more advanced than ELIZA from a conversational standpoint.

In 1981, the first incarnation of the chatbot Jabberwacky was introduced by British programmer, Rollo Carpenter. Carpenter intended for Jabberwacky to simulate natural chat in an interesting, humorous and entertaining manner (Warwick and Shah, 2016). Carpenter continued to evolve his chatbot, but things changed once the Internet gave Jabberwacky access to thousands of online interactions from which it could formulate responses (Boiano et al., 2018). Access to these human conversations allowed Jabberwacky to learn new languages, concepts and facts. This is significant in that, according to its creators, Jabberwacky is not based on any artificial intelligence life model (i.e., Neural Networks, Fuzzy Logic) and is instead a purely heuristics-based technology that relies entirely on context and feedback instead of rules (Bhagwat, 2018). After several second and third place finishes, Jabberwacky variants won the Loebner Prize in 2005 and 2006.

Another multi-Loebner award winner is Richard Wallace's A.L.I.C.E. (Artificial Linguistic Internet Computer Entity), winning the Bronze medal in 2000, 2001, and 2004 (Thompson, 2002; Wallace, 2009). Wallace first developed A.L.I.C.E. to be an artificial intelligence natural language chatbot in 1995. In 1998, A.L.I.C.E. was migrated to the JAVA-platform facilitating platform-independence. The Artificial Intelligence Markup Language (AIML) was developed using an XML-like syntax to define the heuristic conversation rules. A.L.I.C.E. is also an open source chatbot whose source code is available on Google Code and from Richard Wallace's GitHub account. Spike Jonze, the director of the Academy Award winning film *Her* has stated A.L.I.C.E. served as the inspiration for the AI in the film (Morais, 2013).

SmarterChild was born in 2001 as a demo for ActiveBuddy Inc.'s interactive agent platform (Kerr, 2005). SmarterChild operated on Instant Messenger (IM) networks like AOL Instant Messenger (AIM), ICQ, and MSN Messenger and facilitated chatbot communication with IM users in real-time. SmarterChild could perform simple tasks like explaining financial products, predicting the weather, or engaging in humor. SmarterChild could not contact users unsolicited, so users would add them to their "Buddy Lists" just like a human contact. SmarterChild is one of the earliest examples of a "companion bot."

From 2006–2011, IBM developed Watson, a question answering computing system initially developed to answer questions on the quiz show *Jeopardy!* To accomplish this goal, Watson was designed to apply advanced natural language processing, information retrieval, knowledge representation, automated reasoning, and machine learning technologies (Deshpande et al., 2017). In 2011, Watson participated in the *Jeopardy! Challenge* and defeated legendary champions Brad Rutter and Ken Jennings

(Markoff, 2011). Since 2011, IBM has developed many applications on the Watson platform across multiple domains including healthcare, teaching assistants (Leopold, 2017), weather forecasting (Jancer, 2016), tax preparation (Moscaritolo, 2017), and a chatbot providing conversation for children's toys (Takahashi, 2015).

Apple released Siri in 2011, the first of a wave of intelligent personal assistants including Microsoft's Cortana, Amazon's Alexa, and the Google Assistant. Intelligent personal assistants were initially deployed on mobile phones, but have since been developed to operate on personal computers and in-home devices passively listening for user questions or commands. Intelligent personal assistants use natural language processing to give the user access to music services, agenda, news, weather, To-Do lists, maps and directions, and more (López et al., 2017; Sarikaya, 2017).

In 2014, Microsoft released the chatbot XiaoIce in China (Shum et al., 2018). XiaoIce impersonates a 17-year-old Chinese girl and has become a celebrity having been used by over 660 million users across more than 40 platforms, including: WeChat, QQ, Weibo and Meipai in China, Facebook Messenger in USA and India, and LINE in Japan and Indonesia (Zhou et al., 2018). Similar to intelligent personal assistants such as Siri or Google Assistant, XiaoIce can perform tasks such as Web search, weather report, play songs, or provide news recommendations. What has really set XiaoIce apart is its core chat functionality. XiaoIce chats like a Chinese teenager. By focusing on specific empathy models, XiaoIce becomes a confidant and friend to the teenagers who use it. XiaoIce also has a Japanese derivative called Rinna which was launched in 2015 and Ruuh launched in India in 2017.

In 2016, Microsoft released a similar bot named Tay for a Western audience on the Twitter, Kik, and GroupMe platforms. Tay was built to mimic a 19-year-old American girl and learn from its interactions on these platforms (Bright, 2016). However, anonymous users of the notorious troll sites 4chan and 8chan, were able to quickly identify some weaknesses in Tay's "repeat after me" features and exploit them. Tay learned and began tweeting racist, misogynist, and other offensive comments within sixteen hours of its release. Microsoft quickly pulled the plug and deleted the most offensive tweets (Ohlheiser, 2016).

Six months later, Microsoft released another bot for Western audiences named Zo, onto the Kik and Facebook Messenger platforms. Zo mimics a 22-year-old girl who interacts with its users, mostly teens, for an average of about 10 hours of conversation. All of these interactions serve to improve Cortana and Microsoft's bot platforms (Zo, 2019). Zo was shut down in April 2019.

Steve Worswick's Mitsuku has won the Loebner Prize bronze medal in 2013, 2016, 2017, 2018, and 2019 (AISB, 2019). Mitsuku was developed to imitate an 18-year-old female from England and operates on Facebook Messenger, Twitch group chat, Telegram and Kik Messenger under the username "Pandorabots" (AI Dreams, 2013). Mitsuku is a rule-based bot containing all of A.L.I.C.E.'s AIML files.

The Artificial Intelligence Markup Language (AIML) was developed by Richard Wallace from 1995 to 2002 for developing A.L.I.C.E. AIML is an XML-based data structure for representing conversation and dialog primitives. Currently, chatbots using an AIML knowledge base are the most successful in competitions like the Loebner Prize, as shown by the wins by Mitsuku (5 wins), A.L.I.C.E. (3 wins), and Jabberwacky (2 wins) (Bush and Wallace, 2001).

Chapter 5
The Nature of Expertise

In the cognitive systems era, systems will be capable of expert-level performance in virtually any domain. These "cogs" as we call them, will be made available to billions of people across the world via a host of apps and services offered largely via the Internet and used on handheld devices. But what does it mean to be an expert? What is expertise? What do we mean when we say cogs will achieve expert-level performance?

The nature of intelligence and expertise has been debated for decades. As Gobet recently pointed out, traditional definitions of expertise rely on *knowledge* (the know-that's) and *skills* (the know-how's) (Gobet, 2016). A typical traditional definition might read as follows:

"a person having special skill or knowledge in some particular field."

This is a satisfying definition because we expect an expert to know more about a topic and be able to perform skills related to that topic better than the average person. However, Gobet gives a definition of expertise from a different perspective:

"...an expert is somebody who obtains results vastly superior to those obtained by the majority of the population."

This definition focuses more on outcomes achieved rather than the constituent parts of an expert as does the former definition. This point of view has woven through the field of artificial intelligence for decades.

5.1 The Turing Test and Searle's Chinese Room

Turing (1950) described the "imitation game" in which the output of a machine is compared to the output of a human and when an observer is unable to determine which is human and which is machine, the machine can be declared as "intelligent." Put into various forms throughout the decades, this is usually referred to as the "Turing Test."

The notion has been debated over the years. For example, the thought experiment known as Searle's Chinese Room envisions a human within a closed room translating English to Chinese and Chinese to English simply by following a set of rules printed in a book (Searle, 1980). The person is not doing the translation, and in fact, may not even be able to read English or Chinese. The person is simply following the prescribed rules. Messages in one language, slid under the door, are translated and slid back out under the door. If we just look at the inputs and outputs of Searle's Chinese Room, it satisfies the Turing Test. However it begs the question: where is the intelligence?

The Turing Test focuses on the outputs, not how the outputs are produced. Saying the entire room is intelligent does not seem correct. It is not the room itself that is special here but the contents of the room is where the magic happens. Saying the human translator is intelligent is not correct either since they know nothing of English or Chinese. Simply following rules and manipulating symbols is not considered intelligence. Even if it were, it is not intelligence as related to English/Chinese translation.

Is the book of rules therefore where the intelligence lies? Few would be comfortable with saying a static, inanimate object like a book was intelligent. It does contain the knowledge needed for the expertise of translation, but the book is not able to perform actions, so possesses no skills.

What about the book plus the human? The book represents the knowledge and the human represents the skills (the ability to perform actions following the rules). Arguments could be made to this interpretation, and have been over the years. Interestingly, this interpretation almost gets us back to our human/cog idea of synthetic expertise described in this book. The human is a biological entity and the book is an artificial entity. The difference though is the book will never be able to perform any cognitive processing tasks at all, much less cognitive processing approaching the human expert level. In our view of the cognitive systems era, humans will partner with artificial entities capable of performing expert-level cognitive processing. So, it is closer to correct in thinking of human/cog collaboration being partners of two sub-experts who achieve expertise when working together. We later argue human/cog collaboration will lead to a future in which more of the average population will attain expert-level performance—*synthetic expertise* (Fulbright, 2020). More on this later as it is the theme of the book.

5.2 Chunks

Nobel laureate and Turing Award recipient Herbert A. Simon studied the nature of expertise as early as the 1950s. Simon and Gilmartin (1973) estimated experts need to learn on the order of 50,000 "chunks" of information/knowledge about the specific domain. Furthermore, experts gather this store of knowledge from years of experience (on the order of 10,000 hours). This represents the "knowledge" portion of the traditional definition of expertise. Simon and Gilmartin created the Memory-Aided Pattern Perceiver (MAPP) model describing how experts look at a current situation and match it to the enormous store of domain knowledge. (Luc Steels, 1990) later described this type of knowledge as *deep domain knowledge*. In their study of expert Chess players (Chase and Simon, 1973), noted:

> *"... a major component of expertise is the ability to recognize a very large number of specific relevant cues when they are present in any situation, and then to retrieve from memory information about what to do when those particular cues are noticed. Because of this knowledge and recognition capability, experts can respond to new situations very rapidly—and usually with considerable accuracy."*

This view of expertise is based on pattern recognition. Experts match the patterns in the current situation to instances where these patterns have been experienced before. Having found a match in memory, experts retrieve information and solutions from previous experience and then apply those to the current situation. Experts extract from memory much more knowledge, both implicit and explicit, than novices. Furthermore, an expert applies this greater knowledge more efficiently and quickly to the situation at hand than a novice. Therefore, experts are better and more efficient problem solvers than novices. This represents the "skills" portion of the traditional definition of expertise.

The pattern recognition characteristic of expertise is interesting because deep learning and convolutional neural networks are perfect for this kind of processing. As further discussed in Chapter 2 and Chapter 4, deep neural networks detect patterns in data. A stimulus is presented to an artificial neural network, which is allowed to respond. Here the "stimulus" represents the current situation encountered by an expert. Refining and recording a collection of responses is analogous to an expert compiling his or her collection of deep domain knowledge. When the neural network responds in a way similar to a previous response, the neural network has found a match with a previously encountered situation. Pattern recognition such as this is behind many recent achievements of cognitive systems. Classification tasks, such as medical diagnoses, is an exercise in pattern matching and is question answering and knowledge retrieval.

The ability of an expert to quickly jump to the correct solution has been called *intuition*. A great deal of effort has gone into defining and studying *intuition*. Dreyfus argued intuition is a holistic human property which cannot be captured by a computer program (Dreyfus, 1972;

Dreyfus and Dreyfus, 1988). However, Simon et al. argued intuition is just efficient matching and retrieval of "chunks" of knowledge and know-how (Gobet and Simon, 2000; Gobet and Chassy, 2009). The debate between Simon and the Dreyfus brothers' interpretations has persisted for decades.

In this book, we side with Simon. Intuitive human experts seem to be able to jump straight to a solution without any obvious thinking being necessary. However, what is really happening is the expert is matching and retrieving solutions they already know will work in a given situation based on their experience. So, while it is true there is less high-level thinking involved, there is no magic here. Intuition is just highly-efficient pattern matching and recall. Today's cognitive systems are already doing the pattern-matching portion of this and achieving results exceeding humans in many domains. We expect the near future will see cognitive systems able to store and retrieve experiences and solutions and mimic intuition.

The idea of a "chunk" of information can be traced back to Miller, a psychologist studying human memory, who famously realized humans can remember seven plus or minus two things (Miller, 1956). This is why social security numbers and telephone numbers are expressed in groups like they are. However, defining what constitutes a "chunk" of information has been elusive over the decades.

de Groot established the importance of *perception* in expertise in that an expert perceives the most important aspects of a situation faster than novices (de Groot, 1965). Chase and Simon (1973) showed expert chess players recognize board positions and identify the best next move far quicker than novices. In this context, a "chunk" is various positions of pieces encountered before and the sequence of moves one should use in that situation. Simon and Feiganbaum created the EPAM (Elementary Perceiver and Memorizer) to model human learning and concept formation and established the concept of a chunk in cognitive science (Feiganbaum and Simon, 1984). Most cognitive architectures, such as the Soar architecture, described in Chapter 6, include chunks as a feature of knowledge storage and recall (Laird et al., 1986). In Soar, a chunk is a production rule, or set of rules, derived from experiences capturing a piece of knowledge.

Capturing and representing knowledge has been a challenge for artificial intelligence researchers since its beginning. Newell and Simon formulated the notion of symbol manipulation early as the Physical Symbol System hypothesis (PSS) which maintains knowledge can be represented by symbols which can be combined into structured expressions such as predicate statements and production rules (Newell and Simon, 1976).

Various other mechanisms have been created to capture knowledge. Minsky famously created the concept of a *frame* to capture and represent knowledge (Minsky, 1977). A frame is simply a collection of named fields, called *slots*, with values stored in the slots. A frame, or even a collection of related frames, can be considered a *chunk* of knowledge.

Gobet defined a chunk as "a collection of elements having strong associations with one another, but weak associations with elements within other chunks" (Gobet et al., 2001). Gobet has argued the traditional notion of a "chunk" is too simple and too static for real-world situations. Instead, Gobet introduces a *template* as a dynamic chunk with static components and variable, or dynamic components, resembling a complex data structure. A template is in many ways similar to Minky's frames. Templates allow an expert to quickly process information at different levels of abstraction yielding the extreme performance consistent with intuition.

The Chunk Hierarchy and Retrieval Structures (CHREST) cognitive architecture allows a system to *learn* by representing knowledge as network of interconnected nodes where each node is a dynamic chunk (a template). This type of knowledge storage differs from other cognitive architectures using production rules and symbols to represent and store knowledge.

In our Model of Expertise described in Chapter 7, *models* are primary resources allowing the cog to recognize conditions and situations of interest. We envision a model to be an collection of CHREST-like templates. Tasks and problem-solving methods are also primary resources in our model, so a chunk in our model is a collection of models, tasks, and methods retrieved from memory.

5.3 Types of Knowledge and the Knowledge Level

Researchers have developed several ways to represent intelligence and intelligent agents. As shown in Fig. 5-1, Allen Newell recognized computer systems are described at many different levels and defined the *Knowledge Level* as a means to analyze intelligent agents at an abstract level (Newell, 1982).

The lower levels represent physical elements from which the system is constructed. In describing computer systems, the lower levels are based on electronic devices and circuits. The higher levels represent the logical elements of the system.

In general, a level "abstracts away" the details of the lower level. For example, consider a computer programmer writing a line of code storing a value into a variable. The programmer is operating at the Program/ Symbol Level and never thinks about how the value is stored in registers and ultimately is physically realized as voltage potentials in electronic circuits. The details of the physical levels are abstracted away at the Program Level.

Level   **Medium**

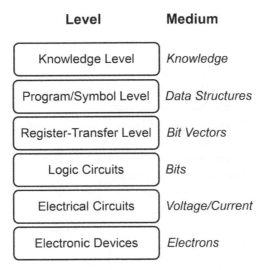

Fig. 5-1: Newell's knowledge level.

Likewise, at the Knowledge Level, the implementation details of how knowledge is represented in computer programs is abstracted away. This allows us to talk about knowledge in implementation-independent terms thus facilitating generic analysis about intelligent agents. An expert is a kind of intelligent agent, so the Knowledge Level can be used to describe experts and expertise as well.

In addition to knowledge required of experts by Simon and earlier researchers, Steels (1990) identified the following as needed by experts: *deep domain knowledge, problem-solving methods,* and *task models.* Problem-solving methods are how one goes about solving a problem. There are generic problem solutions applicable to almost every domain of discourse such as "how to average a list of numbers." However, there are also *domain-specific problem-solving methods* applicable to only a specific or very small collection of domains.

A task model is knowledge about how to do something. For example, how to remove a faucet is a task an expert plumber would know. As with problem solutions, there are generic tasks and domain-specific tasks. Steels' deep domain knowledge represents the "knowledge" in the traditional definition of expertise and the problem-solving and task models represent the "skills" in the traditional definition of expertise. Steels separates problem-solving as a different kind of skill from performing a task.

Steels (1990) also proposed a *Knowledge-Use Level* between Newell's Knowledge Level and the Program Level, as shown in Fig. 5-2, to address issues like task decomposition, execution, scheduling, software architecture, and data structure design.

Level   Medium

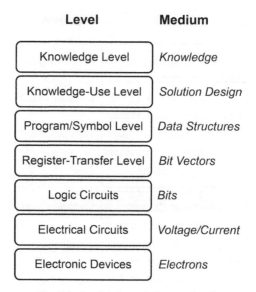

Fig. 5-2: Steele's knowledge-use level.

Whereas the Knowledge Level is implementation independent, the Knowledge-Use Level is geared toward implementation and is quite dependent on implementation details. Steeles felt this level is necessary to bridge the gap between the Knowledge Level and the Program/Symbol Level. Later, we introduce a new level *above* the Knowledge Level called the *Expertise Level*.

5.4 Bloom's Taxonomy

Human cognition has been studied for decades and different levels of cognition have been defined. Bloom's Taxonomy is a famous hierarchy from the field of educational pedagogy relating different levels of cognition as shown in Fig. 5-3. Originally published in the 1950s, and revised in the 1990s, Bloom's Taxonomy was developed as a common language about learning and educational goals to aid in the development of course materials and assessments (Bloom et al., 1956; Anderson et al., 2001).

Bloom's Taxonomy addresses the question of demonstrating mastery of a subject. In the education field, instructors design assessments requiring students to demonstrate the skills at each level of the taxonomy. For example, some exam questions may simply be information retrieval questions corresponding to the *remember* level. Other exam questions and assignments correspond to each of the higher levels in the taxonomy. A student able to exhibit competency at all levels of Bloom's Taxonomy has

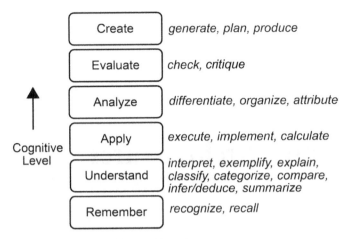

Fig. 5-3: Bloom's taxonomy.

demonstrated a mastery of the subject matter. It occurs to us demonstrating mastery of a certain subject matter is a qualification, and indeed, a definition of an *expert*.

Each level in Bloom's Taxonomy represents a cognitive process with the amount of effort (cognitive work) required increasing dramatically as one goes from *remember* to *create*. The processes are listed in order from the simplest (remember) to the most complex (create). A system, artificial or natural, performing any of these processes is performing some degree of cognition. Currently, our computers and electronic devices perform only the lowest-level processes. In the coming cog era, cogs will come to perform more of, and indeed, all of the processes at every level. Until such time as an artificial system can perform at every level, synthetic expertise will be achieved by a human/cog ensemble, who together, performs at every level of Bloom's Taxonomy. If a human/cog ensemble is performing at every level in Bloom's Taxonomy, then the human/cog is achieving *expertise*. Since some of the cognitive processing is done by an artificial entity, the cog, then we say the ensemble is achieving *synthetic expertise*.

Chapter 6

Cognitive Architectures

An entity capable of intelligent behavior is called an *intelligent agent* but not all intelligent agents perform intelligent behavior (King, 1995). The range of intelligence exhibited by an intelligent agent is large. For example, a thermostat could be considered to be an intelligent agent (Russell and Norvig, 2009).

Intelligent agents are assumed to be situated in an environment which the agent can sense, through various sensors, and alter, by using various effectors to perform actions. Physical disturbances in the environment are sensed and converted into information by the agent. The agent uses various internal knowledge and reasoning mechanisms to decide on an action to take. Actions are performed by manipulating and controlling effectors which change the environment in some way. This *perceive-reason-act* cycle is pervasive throughout cognitive architectures. This, an intelligent agent is anything able to sense and change the environment, even if it does not involve intelligence.

Since the beginning of artificial intelligence as a discipline in the 1950s, researchers have sought to understand intelligence and has constructed a number of formulations to describe and explain intelligent behavior. In general, these formulations are called *cognitive architectures*. Cognitive architectures identify and explain structures in biological or artificial intelligent agents and explore how these structures work together along with knowledge and skills to result in intelligent behavior. This chapter contains cognitive architectures collected from artificial intelligence, cognitive science, and agent theory.

6.1 Genealogy of Cognitive Architectures

Most of the cognitive architectures presented in this chapter come from three generations of researchers who have associations with one another and with universities such as: Carnegie Mellon, University of Michigan,

MIT, and Stanford. Allen Newell (Carnegie Mellon University), Herbert A. Simon (Carnegie Mellon University), John McCarthy (MIT), Marvin Minsky (MIT) and Arthur Samuel (IBM) and a few others are credited with founding and leading the field of artificial intelligence research (McCarthy et al., 1955).

Among Herbert A. Simon's doctoral students were Edward Feigenbaum and Allen Newell. Feigenbaum, is often referred to as the father of expert systems and along with Simon developed EPAM (Elementary Perceiver and Memorizer) a computer program simulating phenomena in verbal learning (Feigenbaum and Simon, 1984). EPAM formalized the notion of a "chunk" and later influenced the CHREST (Chunk Hierarchy and Retrieval Structures) model developed by Gobet and Lane (Gobet et al., 2001).

Allen Newell mentees included Hans Berliner, Paul Rosenbloom, and John Laird. Berliner is known for developing the HiTech chess program and the BKG backgammon program (Berliner, 1977; 1980). Berliner's doctoral student and HiTech team member, Murray Campbell, went on to work on IBM's DeepThought and DeepBlue system which defeated the human chess champion Gary Kasparov in 1997, a landmark event in artificial intelligence. Rosenbloom, Laird, and Newell developed the Soar cognitive architecture, one of the leading cognitive architectures (Laird, 2012).

Laird started work on Soar at Carnegie Mellon but later took a position at the University of Michigan. Developers of the EPIC cognitive architecture, David Meyer and David Kieras graduated from Michigan and are also professors there. Together with Soar, ACT-R, and CLARION, EPIC is considered among the leading cognitive architectures (Kieras and Meyer, 1997).

John Anderson received his PhD from Stanford University and, along with his doctoral student Christian Lebiere, developed the ACT-R (Adaptive Control of Thought-Rational) architecture at Carnegie Mellon (Anderson, 2013). Also associated with Stanford University was Nils Nilson and Michael Genesereth. Nilson and Genesereth authored *The Logical Foundations of Artificial Intelligence* in 1976 which became one of the most read books in artificial intelligence (Genesereth and Nilsson, 1987). One of Genesereth's doctoral students was Stuart Russell who later, with Peter Norvig, wrote *Artificial Intelligence: A Modern Approach* which has become the leading book in artificial intelligence through several revisions (Russell and Norvig, 2009). Several of the cognitive architectures presented in this chapter come from these two books.

Another student of Genesereth, was Jeffery Rosenchein who became known for work in distributed artificial intelligence which focuses on agent-based behavior (Rosenschein and Genesereth, 1988). Tom

Hinrichs, also a student of Genesereth, together with Ken Forbus from MIT, developed the cognitive companion systems concept (Forbus and Hinrichs, 2006). A cognitive companion shares many characteristics of the collaborating cognitive systems (cogs) we focus on in this book. Another Stanford professor, Pat Langley, developed Icarus, a cognitive architecture for physical agents (Langley and Choi, 2006). Langley has challenged the cognitive systems community with creating artificial entities like lawyers as a way to challenge the research community with a goal (Langley, 2013).

One of the original pioneers of artificial intelligence, Marvin Minsky, from MIT, developed the "society of mind" concept (Minsky, 1986), and later refined this to the Emotion Machine architecture (Minsky, 2007). Several of Minsky's students have made major contributions. Carl Hewitt is known for the actor model—the view of systems as collections of interacting actors (agents) (Hewitt et al., 1973). Patrick Winston, with MIT, wrote *Artificial Intelligence* which became a widely-read book in the 1980s and 1990s through several revisions (Winston, 1977). Luc Steels, identified components of expertise, described earlier in this book (Steels, 1990).

6.2 General Problem Solver

As shown in Fig. 6-1, one of the earliest efforts in artificial intelligence research, the General Problem Solver (GPS) was created in 1959 and was intended to be a universal problem-solver machine (Newell et al., 1959). At a given point in time, the machine exists in any of a set of states called the *problem state space* with one state declared the *goal state* as shown in Fig. 6-1. In each iteration, the machine determines the distance from the current state to the goal state and then selects an operator to perform resulting in the machine moving to a state closer to the goal. The machine iterates until the goal is achieved. GPS was the first computer program separating the knowledge of problems from the strategy of how to solve problems. While it was a necessary first logical step, GPS was never able to solve real-world problems. However, GPS did evolve eventually into the Soar architecture described later.

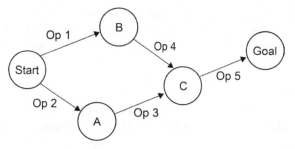

Fig. 6-1: The general problem solver (GPS).

6.3 Reflexive/Tropistic/Instinctive Agent

The simplest architecture to capture the perceive/act nature of agents embedded in an environment is the *reflexive agent* architecture shown in Fig. 6-2 (Russell and Norvig, 2009). A pervasive idea in agent theory is an agent perceives its environment, reasons about it perceptions, and then acts to effect a change in the environment. However, some actions do not require high-level reasoning. For example, if you touch a hot surface with your finger, you will jerk your finger away before your brain has a chance to process the event. In humans, this is called a reflexive action. Reflexive agents are also called *tropistic* or *instinctive* agents (Genesereth and Nillson, 1987).

A reflexive agent perceives the environment and instinctively or reflexively acts without first reasoning about the perception or action. The set of actions the agent can perform is *A*. Each action, when taken, causes the environment to change to another state in S $(A \times S \rightarrow S)$. During each cycle, the environment exists in a state (s $\subset$ *S*) however the agent is not able to detect every state in *S*. The agent can perceive only one of a set of partitions (groups of states) in *S* called *T*. The reflexive agent perceives the environment (*see* function) and detects which partition (t $\subset$ *T*) the environment is in $(S \rightarrow T)$. Having perceived the environment to the best of its ability, the agent selects an action (a $\subset$ *A*) (*action* function) and then performs the selected action (*do* function) causing the environment to transition to a new state $(A \times S \rightarrow S)$.

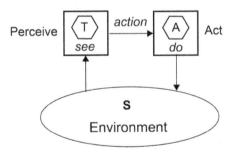

$$\{S, T, A, \text{see, do, action}\}$$

S	Set of states for the environment
T	Partitions of *S* distinguishable by the agent
A	Set of actions the agent can perform

see	$S \rightarrow T$
do	$A \times S \rightarrow S$
action	$T \rightarrow A$

Fig. 6-2: Reflexive/Tropistic/Instinctive agent.

6.4 Model-Based Reflexive/Hysteretic Agent

As shown in Fig. 6-3, a *model-based reflexive* agent (also called a *hysteretic agent*) agent maintains a set of internal representations, *M*, of itself and the environment and incorporates this information into selecting is actions (Russell and Norvig, 2009; Genesereth and Nillson, 1987). The set of actions the agent can perform is *A*. Each action, when taken, (*do* function) causes the environment to change to another state ($A \times S \to S$). During each cycle, the environment exists in a state (s ⊂ *S*) however the agent is not able to detect every state in *S*. The agent can perceive only one of a set of partitions (groups of states) in *S* called *T*. The agent perceives the environment (*see* function) and detects which partition (t ⊂ *T*) the environment is in ($S \to T$). The agent modifies its internal representation (*model* function) based on its new perceptions ($M \times T \to M$). Having perceived the environment and updated its internal model, the agent selects an action (a ⊂ *A*) (*action* function) using both the internal model and its perceptions ($M \times T \to A$). The agent then performs the selected action (*do* function) causing the environment to transition to a new state ($A \times S \to S$).

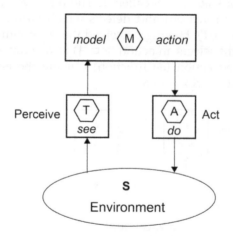

{*M, S, T, A,* see, do, action, model}

M	Set of internal states/models
S	Set of states for the environment
T	Partitions of *S* distinguishable by the agent
A	Set of actions the agent can perform

see	*S → T*
do	*A* x *S → S*
action	*M* x *T → A*
model	*M* x *T → M*

Fig. 6-3: A model-based reflexive/hysteretic agent.

A model is any kind of representation depicting a certain state or condition of interest to an agent. One of the first types of models was Marvin Minsky's *frames* (Minsky, 1977). Frames were one of the first attempts in artificial intelligence research to capture knowledge in an independent form allowing it to be processed (other than in symbols and production rules). Frames consists of a number of *slots* with each slot containing a piece of information about the frame (and even other frames). An example of a frame is shown in Fig. 6-4.

A more dynamic form of a frame, called a template, was developed by Gobet (Gobet et al., 2001; Gobet, 2016). Models allow agents to recognize and classify objects, situations, and conditions in the environment. As an agent perceives the environment, it may "fill in" information about an item. When enough partial information is obtained (enough of the frame's slots are full of information), the agent can conclude knowledge about the item. This feature allows model-based agents to operate in a real-world environment and act robustly with only partial or incomplete information.

For example, an agent may maintain a set of models describing various dangers to be avoided. As time progresses, the agent acquires information about the environment. When enough information is collected in one of the "danger" models, the agent may conclude with some amount of confidence a danger of that type exists and then act accordingly.

Slot	Value	Type
GIRL		(This Frame)
ISA	Person	(Parent Frame)
SEX	Female	(Instance Value)
AGE	<16	(Constraint)
NUM_LEGS	Default=2	(Inherited)
NUM_ARMS	Default=2	(Inherited)

Fig. 6-4: A frame data structure.

6.5 Goal-Based/Knowledge-Level Agent (Reasoning Agent)

As shown in Fig. 6-5, a *goal-based/knowledge-level agent* maintains a store of general knowledge, K. The agent also maintains a set of models, M, the internal representations, and G, a set of goals the agent is trying to achieve (Russell and Norvig, 2009; Genesereth and Nillson, 1987). The agent uses the knowledge, models, and goals to reason about the next action to perform.

The set of actions the agent can perform is A. Each action, when taken, (*do* function) causes the environment to change to another state ($A \times S \rightarrow S$). During each cycle, the environment exists in a state ($s \subset S$) however

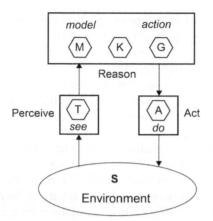

{**K, M, G, S, T, A,** see, do, action, model}

K Set of general knowledge statements
M Set of internal states/models
G Set of goals the agent is trying to achieve
S Set of states for the environment
T Partitions of **S** distinguishable by the agent
A Set of actions the agent can perform

see $S \rightarrow T$
do $A \times S \rightarrow S$
action $M \times G \times K \times T \rightarrow A$
model $M \times G \times K \times T \rightarrow M$

Fig. 6-5: A goal-based/knowledge-level agent.

the agent can detect only one of a set of partitions (groups of states) in S called T. The agent perceives the environment (*see* function) and detects which partition ($t \subset T$) the environment is in ($S \rightarrow T$). The agent modifies its internal representation (*model* function) based on its new perceptions, its knowledge of the world, and its goals ($M \times G \times K \times T \rightarrow M$). Having perceived the environment and updated its internal model, the agent selects an action ($a \subset A$) (*action* function) using the internal models, goals, knowledge, and its perceptions ($M \times G \times K \times T \rightarrow A$). The agent then performs the selected action (*do* function) causing the environment to transition to a new state ($A \times S \rightarrow S$).

Here, general knowledge, K, and goals, G, are logically separated from the models, M. One could use the same data structure to represent all three if desired, but this is not necessarily the case in every implementation. In more sophisticated models, discussed later, different kinds of knowledge will be separated out into independent structures. It is important to note this type of agent is not able to modify its knowledge or goals (it is not a learning agent), but it can modify/update its set of models by incorporating information from recent perceptions.

In general, *M* captures representations about the dynamic nature of the environment while *K* captures static, always-true, concepts. For example, *K* might contain the knowledge "unsupported objects will fall downward in the presence of a gravitational field." This piece of information is true no matter where objects in the environment are positioned so this is a static piece of knowledge about the world. However, objects can move around in the environment, so *M* is used to capture location of objects in real-time as the agent is able to perceive their locations and positions. If the agent perceives an object moved to a position where it is no longer supported, combining that perception with *K* allows the agent to conclude the object will begin to fall downward. The fact the object is going to fall downward is a new piece of knowledge about the environment and allows the agent to choose actions accordingly. For example, if that situation is undesirable, the agent could choose an action to mitigate the object from hitting the floor.

The more general information (common sense knowledge) an agent knows about itself, its tasks, and the environment, the better it can function in real world scenarios. How many pieces of information does an agent need in *K* to be able to handle all of the uncertainty of operating in the real world? This is a famous unsolved problem in artificial intelligence research. No one knows how many pieces of information are required and no one knows if it is even possible to capture all relevant knowledge to facilitate common sense reasoning. Projects such a Cyc (Lenat and Guha, 1989), Open Mind Common Sense and ConceptNet (Havasi et al., 2007), and various projects at the Allen Institute for Artificial Intelligence (Allen Institute, 2019) are seeking to capture volumes of common sense knowledge and bring common sense reasoning to reality, but this is still a work in progress.

An agent possessing substantive *K* is much more robust than an agent without *K*. For example, in the example given above, just one piece of knowledge allows the agent to recognize the problem of an object falling because of the location of an object. More pieces of knowledge might have enabled the agent to prevent the object from being placed in a precarious position in the first place. Still more knowledge might have avoided the entire situation by alternative action.

6.6 Utility-Based/Knowledge-Level Agent (Rational Agent)

As shown in Fig. 6-6, a *utility-based/knowledge-level agent*, also called a *rational agent*, maintains a store of general knowledge, *K*, a set of models, *M* (internal representations), a set of goals the agent is trying to achieve, *G*, and a set of utility values, *U*. The agent uses the knowledge, models, and goals to reason about the next action to perform (Russell and Norvig, 2009; Genesereth and Nillson, 1987).

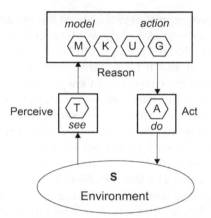

{*K, M, U, G, S, T, A, see, do, action, model*}

K	Set of general knowledge statements
M	Set of internal states/models
U	Set of utility values
G	Set of goals the agent is trying to achieve
S	Set of states for the environment
T	Partitions of **S** distinguishable by the agent
A	Set of actions the agent can perform

see	$S \rightarrow T$
do	$A \times S \rightarrow S$
action	$M \times G \times K \times U \times T \rightarrow A$
model	$M \times G \times K \times U \times T \rightarrow M$

Fig. 6-6: A utility-based agent.

The set of actions the agent can perform is **A.** Each action, when taken (*do* function) causes the environment to change to another state ($A \times S \rightarrow S$). During each cycle, the environment exists in a state (s $\subset$ S) however the agent can detect only one of a set of partitions (groups of states) in S called *T*. The agent perceives the environment (*see* function) and detects which partition (t $\subset$ *T*) the environment is in ($S \rightarrow T$). The agent modifies its internal representation (*model* function) based on its new perceptions, its knowledge of the world, its goals, and its utility values ($M \times G \times K \times U \times T \rightarrow M$). Having perceived the environment and updated its internal model, the agent selects an action (a $\subset$ A) (*action* function) using the internal models, goals, knowledge, utility values, and its perceptions ($M \times G \times K \times U \times T \rightarrow A$). The agent then performs the selected action (*do* function) causing the environment to transition to a new state ($A \times S \rightarrow S$).

The utility values allow priorities to be assigned to items, especially goals and actions. Utility values allow the agent to choose a preferred course of action when it must decide between two or more alternatives. For example, consider two goals with "take pictures of animals" has a lower priority than "collect samples of water." If the agent then finds itself in a situation where it must decide what to do next and both taking pictures and collecting samples are possible next actions, the agent will collect the water sample first.

Utility values also affects the agent's model building performance. For example, if the agent's task is to photograph birds, then bird-objects will have a higher priority than other objects. When the agent perceives a bird, it will record information about the bird into its models. However, if the agent perceives an animal different than a bird, it can decide to ignore it and not update its models with non-bird information.

This type of agent is called a *rational agent* because incorporating utility values into decision processes allows the agent to make decisions yielding the best possible overall performance. When an agent takes the optimal course of action its behavior is considered rational.

Note this agent is not able to change its knowledge, K, goals, G, or utility values, U. That is not to say these values could not be changed by other entities. For example, while the agent is operating, a human operator could assign different utility values thereby altering the agent's behavior. Likewise, another entity could transmit new knowledge to the agent. Once the new knowledge is in K it can be used in all subsequent reasoning and affect the agent's behavior. The next two architectures allow the agent to change its own knowledge, goals, and utility values thereby making them far more autonomous.

6.7 General Learning Agent (Adaptive Agent)

As shown in Fig. 6-7, the set of actions the agent can perform is A. Each action, when taken, (*do* function) causes the environment to change to another state ($A \times S \rightarrow S$). During each cycle, the environment exists in a state ($s \subset S$) however the agent can detect only one of a set of partitions (groups of states) in S called T. The agent perceives the environment (*see* function) and detects which partition ($t \subset T$) the environment is in ($S \rightarrow T$). The agent modifies its internal representation (*model* function) based on its new perceptions, its knowledge of the world, its goals, and its utility values ($M \times G \times K \times U \times T \rightarrow M$). Having perceived the environment and updated its internal model, the agent selects an action ($a \subset A$) (*action* function) using the internal models, goals, knowledge, utility values, and its perceptions ($M \times G \times K \times U \times T \rightarrow A$). The agent then performs the selected action (*do* function) causing the environment to transition to a new state ($A \times S \rightarrow S$). The agent's knowledge, K, can be modified by the agent (*learn* function). Pieces of knowledge can be added, removed, or modified ($M \times G \times K \times U \times T \rightarrow K$). The agent's goals, models, and utility values affect learning.

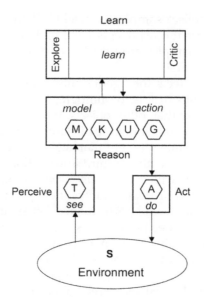

Fig. 6-7: A general learning agent.

{**K, M, U, G, S, T, A,** see, do, action, model, learn}

K   Set of general knowledge statements
M   Set of internal states/models
U   Set of utility values
G   Set of goals the agent is trying to achieve
S   Set of states for the environment
T   Partitions of **S** distinguishable by the agent
A   Set of actions the agent can perform

see $S \rightarrow T$
do $A \times S \rightarrow S$
action $M \times G \times K \times U \times T \rightarrow A$
model $M \times G \times K \times U \times T \rightarrow M$
learn $M \times G \times K \times U \times T \rightarrow K$

The difference between this architecture and the rational agent is the *learn* function. Learning allows the agent's knowledge to evolve over time. Many kinds of machine learning have been developed over the decades in artificial intelligence research but this architecture does not distinguish different types of learning. All forms of learning result in the modification of the knowledge store, *K*.

Allowing the agent's knowledge to change over time affects behavior of the agent since *K* is a factor in the *model* and *action* function as well. For example, new or modified knowledge can change the way the agent builds its models, *M*. Imagine a bird-photographing agent. The agent maintains many models of birds since it is important for the agent to realize a bird is there to be photographed. However, while photographing birds, imagine the agent learns a new piece of knowledge "bears are dangerous" possibly from observing a bear attacking an animal. This new knowledge may cause the agent to create a new model ("bear danger") and start collecting perceived information related to bears. Now, the agent is maintaining bird models and also the bear danger model.

New knowledge can also affect the agent's selection of the next action to perform. Previously, the agent, when sensing a nearby bear, would have ignored the bear. However, with the new knowledge about bears being dangerous, the agent may choose the "flee" or "avoid" actions instead of ignoring the bear.

This is an example of how learning agents adapt to the environment. The ability to adapt is a key skill in real world scenarios where all possible situations cannot be predicted. However, this architecture is not able to autonomously update utility values and goals. Those abilities are found in the next architecture.

6.8 Evolutionary Agent (Evolving Agent)

Figure 6-8 shows an evolutionary (evolving) agent. The set of actions the agent can perform is A. Each action, when taken (*do* function) causes the environment to change to another state ($A \times S \to S$). During each cycle, the environment exists in a state ($s \subset S$) however the agent can detect only one of a set of partitions (groups of states) in S called T. The agent perceives the environment (*see* function) and detects which partition ($t \subset T$) the environment is in ($S \to T$). The agent modifies its internal representation (*model* function) based on its new perceptions, its knowledge of the world, its goals, and its utility values ($M \times G \times K \times U \times T \to M$). Having perceived the environment and updated its internal model, the agent selects an action ($a \subset A$) (*action* function) using the internal models, goals, knowledge, utility values, and its perceptions ($M \times G \times K \times U \times T \to A$). The agent then performs the selected action (*do* function) causing the environment to transition to a new state ($A \times S \to S$). The agent's knowledge, K, can be modified by the agent (*learn* function). Pieces of knowledge can be added, removed, or modified ($M \times G \times K \times U \times T \to K$). The *alter* function allows the agent to change its goals over time ($M \times G \times K \times U \times T \to G$). The *assess* function allows the agent to adjust its utility values ($M \times G \times K \times U \times T \to U$).

The ability for an agent to change its utility values and goals gives it further ability to adapt to real-word situations and achieve the utmost in robust behavior. For example, the goal of "recharge batteries" for a bird-photographing agent would have a fairly low utility value when the battery level is high. In such a condition, the agent will proceed with its normal activity of detecting and photographing birds. However, as the battery level drops, the agent can change the utility value of "recharge

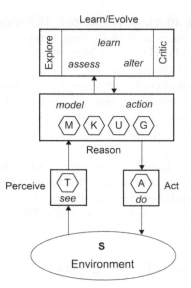

{*K, M, U, G, S, T, A,* see, do, action, model, learn, alter, assess}

K	Set of general knowledge statements
M	Set of internal states/models
U	Set of utility values
G	Set of goals the agent is trying to achieve
S	Set of states for the environment
T	Partitions of **S** distinguishable by the agent
A	Set of actions the agent can perform

see	$S \rightarrow T$
do	$A \times S \rightarrow S$
action	$M \times G \times K \times U \times T \rightarrow A$
model	$M \times G \times K \times U \times T \rightarrow M$
learn	$M \times G \times K \times U \times T \rightarrow K$
alter	$M \times G \times K \times U \times T \rightarrow G$
assess	$M \times G \times K \times U \times T \rightarrow U$

Fig. 6-8: An evolving agent.

batteries" thereby changing the degree of importance to the agent. Changing the utility value to an intermediate level might cause the agent to begin navigating toward a recharging station but also continue with its bird detection and photographing tasks along the way. As the battery level gets critically low, the utility value of "recharge batteries" can be changed to a level making achieving that goal the most important thing to the agent—even surpassing bird detection and photographing. The agent in this condition will stop trying to detect and photograph birds and perform actions only relevant to reaching the recharging station.

Likewise, an agent with the ability to alter its own goals gives the agent a degree of autonomy not available in other architectures. For example, as conditions change for our bird-photographing agent, the agent could establish new goals such as: "photograph bears," "count birds," "classify and count birds," etc. By acquiring new knowledge, setting new goals for itself, and adjusting utility values, the agent evolves new behaviors.

Indeed, this is a picture of a human. When born, a baby possesses a certain set of abilities but immediately begins learning and continuously does so for the rest of its life. Basic goals such as "seek food," "seek comfort," and "seek safety" are established quite early, but humans evolve other goals as time passes such as "graduate from college" and "buy a car." Over time, evolving agents, human or artificial, evolve into individual entities each with a different set of goals, utility values, and knowledge.

In this book, we are concerned with artificial entities, cogs, able to perform at or above human-expert levels. Later, we will argue cogs should be evolving agents as described in this architecture. We will introduce a more detailed architecture—our model of expertise—but at its core, that architecture will be equivalent to the architecture of an evolving agent introduced here.

6.9 EPIC

The cognitive architectures presented so far in this chapter are collectively called the "formal models." The next series of cognitive architectures are based on studying human cognition. We call this group the "humanistic models." The most successful cognitive architecture in this group is the Soar architecture. However, Soar has evolved to incorporate elements from several other architectures including: EPIC, ACT-R, and CLARION.

As shown in Fig. 6-9, the Executive Process-Interactive Control (EPIC) architecture was developed in the 1990s by David E. Kieras and David E. Meyer at the University of Michigan (Kieras and Meyer, 1997). Compare this architecture with the formal models described above. EPIC has the same basic perceive-reason-act structure. The "perceive" function is broken down into auditory, visual, and tactile processing. The "act" function is broken down into vocal, ocular, and manual processing. Since EPIC is derived from studying human cognition, it is not surprising these are the only sensor and effector categories. The perceptual processors perform the "see" function depicted in the formal models and the motor processors perform the "act" function depicted in the formal models.

Reasoning is done in the cognitive processor. In EPIC, all cognition is done by executing production rules. Production rules are stored in the production memory and are retrieved into the rule interpreter when needed. Information needed to execute the production rules comes from two sources: long-term memory and short-term memory. Some information is retained over significant lengths of time in long-term memory. Presumably, this would include general knowledge (K in the

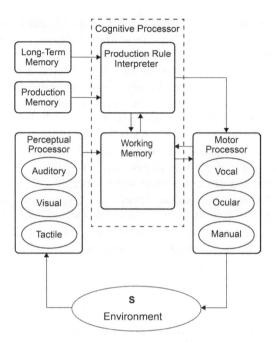

Fig. 6-9: The EPIC cognitive architecture.

formal models) and could include goals (*G*), utility values (*U*) and models (*M*), but these are not expressly represented in the EPIC architecture.

Information in short-term memory (working memory) comes from the perceptual processors and represents the agent's perceptions about the environment (*T* in the formal models). Production rule execution places more information in working memory which is used by the motor processors to carry out actions (*A* in the formal models).

Note, EPIC does not include learning. Knowledge is encoded a priori into the production rules and long-term memory but there is no mechanism for updating this knowledge nor the creation of new production rules.

6.10 ACT-R

As shown in Fig. 6-10, the Adaptive Control of Thought-Rational (ACT-R) architecture was developed by John R. Anderson and Christian Lebiere at Carnegie Mellon University (Anderson, 2013). Work leading to ACT-R began in the 1970s influenced by Allen Newell and the idea of developing unified theory of cognition by studying human cognition.

Like EPIC and the formal models, ACT-R features a basic perceive-reason-act structure. However, ACT-R's perception is depicted as only a visual module. Presumably, other modules could be added as needed to

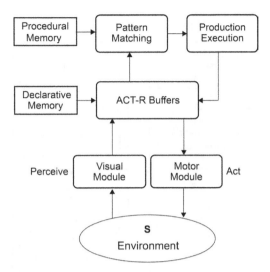

Fig. 6-10: The ACT-R cognitive architecture.

cover other forms of sensory capabilities. The "act" function is represented by the motor module and likewise could be expanded to include other effector elements as needed. Also, like EPIC, ACT-R maintains a set of production rules which feed a pattern matching and production rule execution module. Buffers serve as ACT-R's short term, or working, memory. Information from the perceptual module (*T* in the formal models) is fed into the buffers and used in production rule execution.

ACT-R assumes knowledge is divided into two types of representations: *declarative* and *procedural*. Declarative knowledge (e.g., "Washington, D.C. is the capital of United States") is represented in the form of *chunks* (structured collections of labeled data). Chunks accessible through buffers and may be stored in declarative memory. Procedural memory consists of production rules representing knowledge about how to do things (e.g., how to type the letter "Q" on a keyboard). Together, declarative and procedural knowledge represent the agent's knowledge (*K* in the formal models).

6.11 CLARION

As shown in Fig. 6-11, the Connectionist Learning with Adaptive Rule Induction On-Line (CLARION) cognitive architecture was developed in the early 2000s by a research group led by Ron Sun at the Rensselaer Polytechnic Institute (Sun, 2002).

CLARION makes a distinction between *implicit* and *explicit* knowledge (corresponding to *K* in the formal models). Implicit knowledge is gained

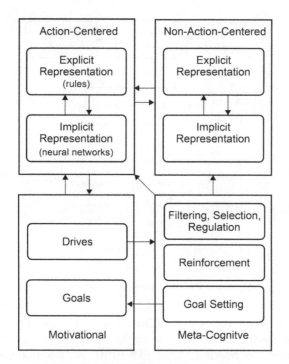

Fig. 6-11: The CLARION cognitive architecture.

through experience and incidental activities (e.g., learning how to ride a bicycle). Explicit knowledge is knowledge easily represented and transmitted to others (e.g., Earth's atmosphere is 21% oxygen). Note, learning is embedded in the architecture so the agent's implicit and explicit knowledge can change over time.

The other distinction is between *action-centered* and *non-action-centered* activities. The Action-Centered Subsystem (ACS) is where all of the agent's actions (both instinctive and learned) are stored (*A* in the formal models). The Non-Action-Centered Subsystem (NACS) maintains general knowledge (*K* in the formal models). Some knowledge is *semantic knowledge* (general knowledge statements about the world) and some is *episodic knowledge* (knowledge about specific situations).

The Motivational Subsystem (MS) is a unique feature of CLARION providing the agent with motivation for perception, cognition, and action. These correspond to goals and utility values in the formal models (*G* and *U*). Drives include low-level motivations (e.g., hunger, avoidance) and high-level motivations (e.g., affiliation, fairness).

Another unique feature is the Meta-Cognition Subsystem (MCS). MCS monitors and directs the operations in the other three subsystems.

Learning is an example of this. Reinforcement learning allows the agent to modify its knowledge. Both actions and non-actions can be learned. The agent can also set/modify its own goals. Therefore, CLARION has elements corresponding to the *alter, assess,* and *learn* functions found in the evolutionary agent formal model.

CLARION is also unique from other architectures in this group because it employs *connectionism.* Connectionism is a branch of artificial intelligence using artificial neural networks (ANNs) to describe and replicate intelligence (McCulloch and Pitts, 1943; Hebb, 1949; Medler, 1998). Connectionism is inspired by the structure of the human brain and is based on interconnected networks of simple units (like neurons in the human brain). When such a network is presented with a stimulus (some form of input information) it causes signals to flow from unit to unit regulated by weighting factors on each connection. The network's response can be fine-tuned by changing the weights on the connections. Over several trials, the response of the network can be tuned to be different for different stimuli. After training, when presented with a similar stimulus, an ANN can determine the kind of stimulus by comparing its response to the responses formed by the training set. This gives connectionist networks a great deal of robustness in dealing with unstructured and partial data which historically have caused production rule systems problems.

6.12 Soar

Figure 6-12 shows the Soar cognitive architecture. Like much of artificial intelligence research, the Soar architecture is based on the idea everything can be represented as symbols and intelligence is the result of symbol manipulation. This is called the Physical Symbol System Hypothesis (PSSH) (Newell and Simon, 1976). Allen Newell was one of the founders of artificial intelligence and the PSSH. As John Laird's and Paul Rosenbloom's dissertation advisor, Newell, Laird, Rosenbloom developed the first version of Soar in the early 1980s (Laird and Newell, 1987). Laird is now the lead of Soar which has been in continual development for over 30 years (Laird, 2012).

Held in long-term memory, Soar maintains a set of procedural production rules (how to do things), a set of semantic representations (knowledge about things), and a set of episodic memory representations (snapshots of short-term working memory). These represent the agent's knowledge (*K* in the formal models).

Soar also features a learning module for each kind of long-term memory store. Reinforcement learning supports procedural memory, semantic learning supports semantic memory, and episodic learning supports episodic memory. This layer in Soar corresponds to the *learn* function in the formal models.

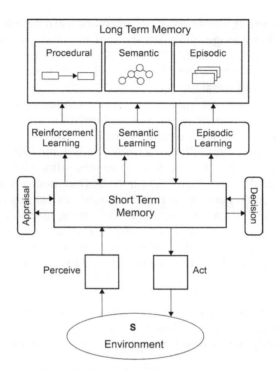

Fig. 6-12: The soar cognitive architecture.

Like other humanistic models, including EPIC and ACT-R described above, Soar makes a distinction between long-term memory and short-term, or working, memory. Information, knowledge, and production rules are brought into short-term memory when needed. While Soar does not provide detail for sensory and effector modules, the agent's perception brings information into short-term memory to provide the agent with information about the environment. Short-term memory is where the production rules are executed using the information and knowledge to reason about the world. The agent's actions are performed based on information from short-term memory. The *decision* and *appraisal* functions support reasoning as well.

At any point in time, short-term memory contains all information of importance at that particular time. Snapshots of working memory are stored in a temporal fashion in episodic memory. At any later time, prior episodes can be retrieved into working memory through. This is how Soar agents are able to recall past experiences. Being able to retrieve knowledge and past experience is key to being an expert as described in later chapters.

6.13 Soar/Formal Model

Here, we combine Soar and the formal models. Soar is the most highly-regarded cognitive architecture developed over a period of more than 30 years (Laird, 2012). The formal models of agents presented earlier in the chapter were developed by artificial intelligence researchers over several decades as well but were developed from a different perspective (Russell and Norvig, 2009; Genesereth and Nillson, 1987). However, the two fundamentally represent the same thing—cognition. Here, we combine the Soar architecture with elements from the formal agent models two as shown in Fig. 6-13. The fit is not perfect but one can see how they complement one another.

The fundamental and classic perceive-reason-act-learn structure is present. All agents are able to perceive a subset of the environment and all agents can perform actions to alter the environment in some way. The formal agent models do not attempt to distinguish long-term from short-term memory. The humanistic models do because these structures appear in humans. The formal models name specific stores such as *goals*, *models*, and *utility values* whereas Soar does not. Soar identifies *kinds* of

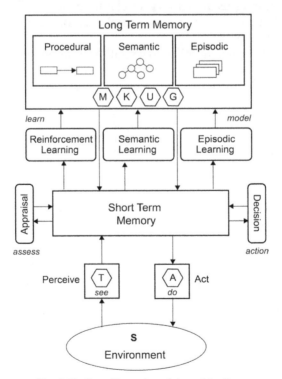

Fig. 6-13: Soar/Formal model combination.

knowledge stores, *procedural, semantic,* and *episodic* because it is intended to explain human cognition. Any formal agent model can be reconciled with Soar by placing various types of knowledge long-term memory. The agent retrieves information and knowledge from these stores and brings them into short-term memory as it does its processing.

Soar also differentiates three different kinds of learning, again, because these kinds of learning are present in human cognition. The formal models do not differentiate and have only a generic *learning* element. However, the formal models could be extended with such detail as desired.

6.14 Standard Model of the Mind

The latest work from the "humanistic models" researchers, Laird, Rosenbloom, and Lebiere, is the Standard Model of the Mind (SMM) shown in Fig. 6-14 (Laird et al., 2017). SMM is seen as a coming together of Soar, Act-R, and several other humanistic cognitive architectures. SMM views the human mind as a collection of independent modules having distinct functionalities and exhibits the classic perceive-reason-act structure. Knowledge is broken up into declarative long-term memory and procedural long-term memory. Declarative memory contains general knowledge about things and procedural memory contains knowledge about how to do things.

SMM is based on symbol representation of knowledge and relations over those symbols. Items stored in memory also include metadata (data about data) such as frequency, recency, co-occurrence, similarity, and

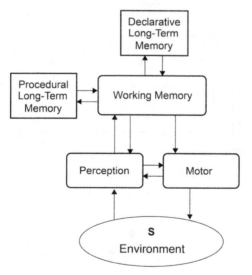

Fig. 6-14: The standard model of the mind.

utility information. This permits statistical treatment of knowledge and a form of episodic memory.

As in EPIC, ACT-R, and Soar, working memory in SMM is where dynamic symbol manipulation occurs. Items from perception and from long-term memory are brought into working memory where inferencing occurs.

Learning in SMM involves the automatic creation of new symbol structures, plus the tuning of metadata, in long-term memory. SMM assumes all types of long-term knowledge are learnable.

6.15 Subsumption Architecture

The next three cognitive architectures form the "subsumptive group." These architectures are interesting because they are based on multiple layers, or levels, which "execute" in parallel. Each successive layer constitutes a higher-level behavior. Overall behavior emerges from the individual actions at each layer.

As shown in Fig. 6-15, the subsumption architecture was created in the 1980s by Rodney Brooks as a reaction to the failures of traditional symbol-based cognitive architectures (the humanistic models described earlier) to achieve success in real-world environments (Brooks, 1986). Systems were "brittle" meaning they could not perform well when exposed to incomplete and sometimes contradictory information in the real-world. Unlike EPIC, ACT-R, CLARION, Soar, and SMM, there is no symbolic representation in the subsumption architecture. Neither is there internal storage of knowledge. Subsumptive agents are not able to store

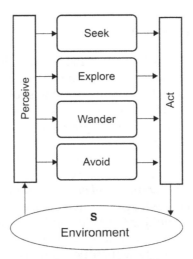

Fig. 6-15: The subsumption architecture.

general knowledge, procedural knowledge, episodic knowledge, nor are they capable of forming models and maintaining goals or utility values. Instead, each layer of behavior is autonomous (although there is feedback between the layers). All behavioral layers operate at the same time and a behavior can subsume (override) another behavior.

The architecture was applied to robotics. For example, an autonomous robot could move around an environment and *avoid* obstacles. At this layer, all the agent does is perceive the environment and act so as to avoid collisions. However, at a higher level, the agent is also *wandering*. Wandering implies a non-random movement. However, while wandering around a room, the agent also avoids collisions. For example, if the agent is wandering in a general northernly direction, it may deviate temporarily to avoid a collision and resume the northernly path after the avoidance maneuver is completed. Higher-level behaviors are *explore* and *seek*. However, as the agent performs these higher-level behaviors it still wanders and avoids.

This architecture is based on the Creature Hypothesis because it mimics insect and animal behavior as opposed to the Physical Symbol System Hypothesis common to the humanistic models. Subsumptive behavior yielded demonstrations of robustness. Lab-scale systems exhibited the ability to maintain overall behavior in the presence of unexpected situations and dynamic environments. However, the subsumptive architectures were not able to evolve complex actions such as high-level cognition, reasoning, and learning. It seems both subsumptive and symbolic approaches are needed to create a complete model of intelligence. Over the last twenty years, several other interesting subsumptive architectures have been created.

6.16   Emotion Machine—Model 6

Marvin Minsky created the Society of the Mind architecture in the 1980s, about the same time Brooks was creating the subsumption architecture described above (Minsky, 1986). Similar to the subsumption architecture, Minsky models different levels of thinking as layers. The Emotion Machine architecture shown in Fig. 6-16 is Minsky's latest version of this type of architecture (Minsky, 2007).

Humans possess *instinctive reactions* to perceptions (e.g., hunger, fear) but also *learn reactions* through experience (e.g., don't step in front of a car). At a higher level, the *deliberative thinking* level, humans can reason about their perceptions and knowledge and make informed decisions. Humans also analyze their own thinking (*reflective thinking*) and constantly examine what-if scenarios to improve future deliberative decisions. At higher levels, humans think about how their perceptions, actions, and reasoning

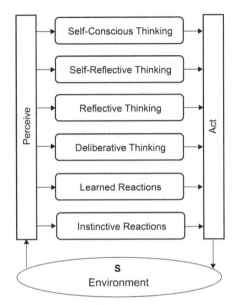

Fig. 6-16: The emotion machine—Model 6.

affects them personally. At the *self-reflective* level (the personal level), humans evaluate themselves against their beliefs, morals, and goals. At the *self-conscious level* (the social level), humans evaluate themselves with respect to others and society in general (e.g., what do others think about my behavior?).

None of these levels are possible without *common sense* and learning. This is called the "Common Sense Hypothesis." Recall, the humanistic models all held general and common sense knowledge in long-term memory stores. The same is presumably true for the Emotion Machine but details are not provided in the architecture. However, as noted earlier, the problem of collecting enough common sense knowledge to be effective is an unsolved problem in artificial intelligence. Projects such a Cyc (Lenat and Guha, 1989), Open Mind Common Sense and ConceptNet (Havasi et al., 2007), and various projects at the Allen Institute for Artificial Intelligence (Allen Institute, 2019) have sought to capture volumes of common-sense knowledge and bring common sense reasoning to reality.

Importantly, the Emotion Machine distinguishes different levels of human thinking and is ultimately a model of the human mind. A still unanswered question is whether or not artificial systems need all of these levels of thinking to be effective. This book involves synthetic expertise—systems capable of expert-level cognition in a given domain—so we expect an artificial expert to possess some of the same levels of thinking as a human expert.

6.17 Genesis Architecture

The Genesis architecture shown in Fig. 6-17 is Patrick Winston's latest subsumptive/level-based description of interaction between intelligent agents or beings. The basis of this architecture is the "Strong Story Hypothesis" maintaining intelligence is the ability to tell stories and understand them (Winston, 2011). With humans, language is the mediator of perceiving the world and understanding it. Humans perceive the world and use language to describe it (general and perceptual knowledge) and partition it and time into events (episodic memory). Sequences of events constitute a story. Stories collected at various hierarchical levels such as the family/personal level (microlevel) and the state/country level (macro) constitute culture.

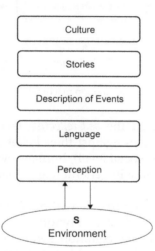

Fig. 6-17: The genesis architecture.

The architecture as shown has the *perceive* function represented but does not represent the *effector* function of an agent. Such details are not the purpose of the architecture but presumably *act* functions would either be done at the same level as the *perception* function. Like the Emotion Machine, the Genesis architecture is of interest to this book because synthetic experts will have to interact naturally with humans. We expect synthetic experts to have the capability of multi-level thinking like that shown in the Emotion Machine architecture and also be able to converse with humans on multiple levels as shown in the Genesis architecture.

Chapter 7

Synthetic Expertise

In this chapter, our Model of Expertise is introduced. The Model of Expertise is implementation independent so can describe both biological and artificial systems. In Chapter 5, basic requirements of an expert are discussed, In Chapter 6, a review of cognitive architectures is presented. Chapters 5 and 6 form the basis of the expertise model introduced in this chapter.

7.1 Requirements for Expertise

Summarizing Chapters 5 and 6, an expert must possess general knowledge, problem-solving skills, and knowledge about how to perform tasks. An expert must have generic problem-solving and task knowledge, applicable to any situation, and also domain-specific problem-solving and task knowledge. Knowledge must be general in nature (knowledge about things), specific in nature (deep domain knowledge), and also common sense in nature. An expert must maintain episodic knowledge as well (knowledge about past experiences). As an evolving agent, an expert must also maintain a collection of models, both generic and domain specific, goals, and utility values. As an intelligent agent, an expert must be able to perceive the environment and recognize a set of environmental configurations. Finally, an expert must have a set of actions able to be performed to effect a change in the environment.

An expert must be able to extract from its knowledge information about what to do in a given situation. The agent's domain-specific models, domain-specific knowledge, and episodic knowledge facilitate recognition of situations. Extraction of problem-solving and task knowledge relevant to the situation completes the "extraction" skills of an expert. An expert must be able to recall, apply, evaluate, understand, analyze, and create with respect to its domain of discourse. We also expect an expert to be able to teach others about the domain.

Finally, learning is inherent in all aspects of an expert. As an expert experiences new situations, knowledge of all types is extracted and used to enhance the agent's knowledge stores. The agent also continually updates its collection of models as well as its problem-solving and task knowledge. As an evolving agent, an expert continually evaluates and updates its goals and utility values.

7.2 The Knowledge Level Description of Expertise

The knowledge an expert must possess is described at the Knowledge Level as shown in Fig. 7-1. Here we combine types of knowledge from the formal models of intelligent agents together with the kinds of knowledge identified by Simon, Steels, and Gobet. Because this representation is at the Knowledge Level, implementation details about how the knowledge stores are realized are not specified nor implied. Refer to Chapters 5 and 6 for details.

An expert is in general an intelligent agent perceiving the environment, reasoning about those perceptions using its internal knowledge, and performing actions thereby causing changes in the environment. Because of limitations in its sensory systems, an expert perceives only a subset

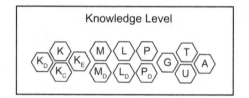

K	declarative knowledge statements (generic)
K$_D$	domain-specific knowledge
K$_C$	common-sense knowledge
K$_E$	episodic knowledge

M	world models (generic)
M$_D$	domain-specific world models
L	task models (generic)
L$_D$	domain-specific task models
P	problem-solving models (generic)
P$_D$	domain-specific problem-solving models

G	goals to achieve
U	utility values
T	perceivable states
A	actions

Fig. 7-1: Knowledge-level description of expertise.

of the possible states of the environment, T. The expert also has a set of actions, A, it can perform to change the environment. Experts are goal-driven and utility-driven evolving agents, as described in Chapter 6, where G represents the set of goals and U represents the set of utility values.

In addition to deep domain knowledge K_D (knowledge about the domain) experts possess general background knowledge K (generic knowledge about things), common-sense knowledge K_C, and episodic knowledge K_E (knowledge from and about experiences). A model, similar to Gobet's *templates* and Minksy's *frames* is an internal representation allowing the expert to classify its perceptions and recognize/differentiate situations. For example, an expert plumber would have an idea of what a leaky faucet looks, sounds, and acts like based on experience. This mental model of a leaky faucet allows the plumber to quickly recognize a leaky faucet when encountered. Some models are domain-specific, M_D, and other models are generic, M. In humans, the collection and depth of models is attained from years of experience. As models are learned from experience, creating and maintaining M_D requires K_E and K_D as a minimum but may also involve other knowledge stores.

As Simon and Steels identified, an expert must know how to solve problems in a generic sense, P, and how to solve problems with domain-specific methods, P_D. These represent the problem-solving knowledge of the expert. In addition, experts must also know how to perform generic tasks, L, and domain-specific tasks, L_D.

An expert is always learning therefore all knowledge types, models, task models, and problem-solving methods are continually changing. Experts can learn both generic and domain-specific forms of new knowledge and also refine knowledge already stored. An expert uses its knowledge to reason about what it perceives, infer and deduce causes and relationships, and consider the effect of possible actions.

However, the Knowledge Level description of expertise is incomplete because it does not include the *skills* an expert must possess. For this purpose, we introduce the *Expertise Level* as shown in Fig. 7-2.

7.3 The Expertise Level

The knowledge and skills an expert must possess are described in Chapter 5. An expert's knowledge is described above at the Knowledge Level as shown in Fig. 7-1. However, a model of expertise must accommodate both expert knowledge and expert skills. We seek a way to represent *skills* in a way not requiring us to worry about the *knowledge* required to perform those skills. Therefore, we introduce a new level called the *Expertise Level* lying above the Knowledge Level as shown in Fig. 7-2. The Expertise Level represents *skills* an expert must possess. At

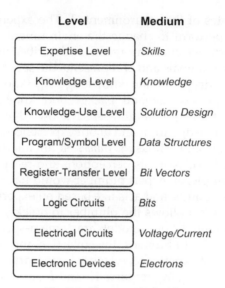

Level	Medium
Expertise Level	*Skills*
Knowledge Level	*Knowledge*
Knowledge-Use Level	*Solution Design*
Program/Symbol Level	*Data Structures*
Register-Transfer Level	*Bit Vectors*
Logic Circuits	*Bits*
Electrical Circuits	*Voltage/Current*
Electronic Devices	*Electrons*

Fig. 7-2: The expertise level.

the Expertise Level, we talk about what an expert does—the skills—and not worry about the details of the knowledge required to perform these skills. Therefore, the medium of the Expertise Level is *skills*.

7.4 The Expertise Level Description of Expertise

Figure 7-3 shows the skills needed by an expert—the Expertise Level description of an expert. Being an intelligent agent operating in an environment necessitates the *perceive* and *act* skills. Being a goal-driven intelligent agent, the *alter* skill allows the expert to change its goals over time. An expert acquires general knowledge, domain-specific knowledge, problem-solving knowledge, and task knowledge over time so the *learn* skill is necessary. Following studies of experts from Simon, Steels, and Gobet, among others, the *extract* skill is the way an expert matches its perceptions about a current situation with its knowledge and retrieves relevant chunks. We add the *teach* skill because we believe any expert should be able to teach someone else about their domain of discourse. To these skills, we add the six skills identified in Bloom's Taxonomy: *recall, understand, apply, analyze, evaluate,* and *create*. Note the *assess* function, allowing an intelligent agent to change utility values, is subsumed by the *evaluate* skill.

The result is twelve fundamental skills needed by an expert: *recall, apply, evaluate, understand, analyze, create, extract, teach, learn, alter, perceive,* and *act*. We call these the *fundamental* skills because any other skill an

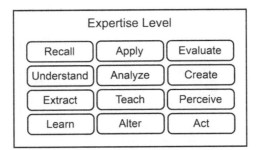

Expertise Level		
Recall	Apply	Evaluate
Understand	Analyze	Create
Extract	Teach	Perceive
Learn	Alter	Act

Perceive	sense/interpret the environment
Act	perform action affecting environment
Recall	remember; store/retrieve knowledge
Understand	classify, categorize, discuss, explain, identify
Apply	implement, solve, use knowledge
Analyze	compare, contrast, experiment
Evaluate	appraise, judge, value, critique
Create	design, construct, develop, synthesize
Extract	match/retrieve deep knowledge
Learn	modify existing knowledge
Teach	convey knowlege/skills to others
Alter	modify goals

Fig. 7-3: Expertise-level description of expertise.

expert may exhibit is a combination of several of these fundamental skills. We talk about higher-level composite skills in a later section.

As described in Chapter 5, the *extract* skill represents the expert's ability to match current perceptions, T, with its episodic and domain knowledge as well as with its models (K_E, K_D, M, M_D) and retrieve chunks of problem-solving, task, episodic, and domain knowledge pertinent to the situation $(P, P_D, L, L_D, K_E, K_D)$. The expert *understands* and *analyzes* what it perceives using common sense and general knowledge (K, K_C) along with what it *recalls* and *extracts* from its knowledge stores. The expert can then *evaluate* the effect of possible actions considering the context of its goals (G) and the weight of its utility values (U). Ultimately, the expert selects an action, A, and *acts* thereby changing the environment and starting the cycle over again. In parallel with this perceive-reason-act cycle, the expert continually *learns* new knowledge, *alters* its goals if necessary, and *evaluates* its utility values considering its recent experience.

7.5 The Model of Expertise

Figure 7-4 is a combination of Figs. 7-1 and 7-3, the Knowledge Level and Expertise Level description of an expert, yielding the formal Model of Expertise.

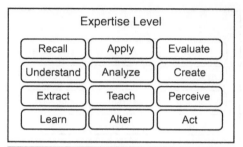

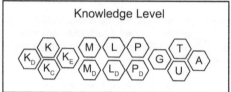

Knowledge
K   declarative knowledge statements
K_D domain-specific knowledge
K_C common-sense knowledge
K_E episodic knowledge   **G** goals to achieve
M/M$_D$ world models   **U** utility values
L/L$_D$ task models   **T** perceivable states
P/P$_D$ problem-solving models **A** actions

Skills
Perceive   sense/interpret the environment
Act   perform action affecting environment
Recall   remember; store/retrieve knowledge
Understand classify, categorize, discuss, explain, identify
Apply   implement, solve, use knowledge
Analyze   compare, contrast, experiment
Evaluate   appraise, judge, value, critique
Create   design, construct, develop, synthesize
Extract   match/retrieve deep knowledge
Learn   modify existing knowledge
Teach   convey knowlege/skills to others
Alter   modify goals

Fig. 7-4: Formal model of expertise.

7.6 The Soar Architecture for Expertise

Having identified the requisite knowledge and skills of an expert, we can now situate those into a Soar-based cognitive architecture as shown in Fig. 7-5. For this, we superimpose our formal model onto the basic Soar architecture of an evolving agent.

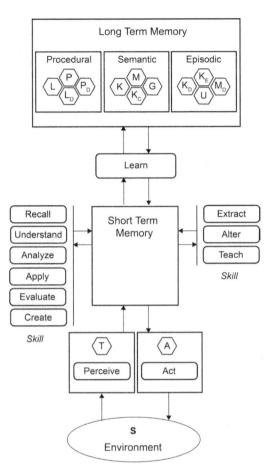

Fig. 7-5: The Soar model of expertise.

7.7 Synthetic Expertise: The Human/Cog Ensemble

Biological systems capable of performing all skills and acquiring/possessing all knowledge required to be an expert, as shown in Fig. 7-4, are *human experts*. Non-biological systems capable of the same are called *artificial experts*. We envision a future in which artificial experts achieve or exceed the performance of human experts in virtually every domain. However, we think it will be some time before fully autonomous artificial experts exist.

In the immediate future, humans and artificial systems will work together to achieve expertise as an ensemble—*synthetic expertise*—as shown in Fig. 7-6. (Compare Fig. 7-6 to Fig. 3-2.) Total cognition of the ensemble is the emergent result of a mixture of biological and artificial thinking. The

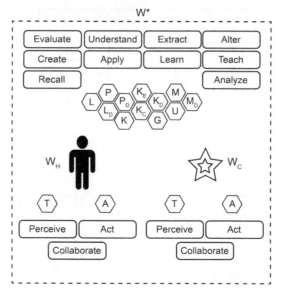

Fig. 7-6: Synthetic expertise—human/cog collaboration.

cognitive work performed by the human is W_H and the cognitive work performed by the cog is W_C. The cognitive work performed by the entire ensemble is W^*.

Together, the human and the cog will perform all skills necessary to be an expert in a domain of discourse as indicated by the fundamental skills shown in Fig. 7-6. Since the human and the cog are physically independent entities, they both perceive the environment and act on the environment. Therefore, the *perceive* and *act* skills along with perceptions T and actions A are associated individually with the human and the cog in the figure.

The other skills and knowledge stores are logically considered to be distributed across the human and cog. Recalling Engelbart's HLAM/T framework from Chapter 3, in some cases, the human will perform all of a specific skill (Engelbart's explicit-human processes). In other cases, the cog will perform all of a specific skill (Engelbart's explicit-artifact processes). Most skills in a human/cog ensemble will be partially performed by the human and partially performed by the cog (Englebart's composite processes). In the figure, the skills are shown as a property of the ensemble and not necessarily a property of the human or the cog individually.

Likewise, the knowledge stores should be considered as a property of the ensemble. Realistically, knowledge will reside within the human mind and knowledge will reside in the cog's electronic memory. This is inescapable given the current physical limitations of brain/computer interfaces (but may not always be true when direct mind-computer

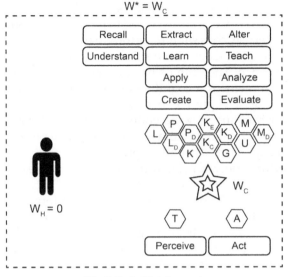

Fig. 7-7: Artificial expert.

interfaces become reality). However, logically, we represent the knowledge stores in the figure to accentuate the collaborative nature of the ensemble.

Eventually, as the capability of cogs increases with future development, the cog will perform all skills required of an expert and we will have developed a truly *artificial expert* as shown in Fig. 7-7 achieving Level 5 cognitive augmentation.

Until we develop stand-alone, autonomous artificial experts, a human component will remain necessary in the ensemble. When the human/cog ensemble performs at a level of, or exceeding, that of an expert, we will have achieved *synthetic expertise* and Level 3 or Level 4 cognitive augmentation. We choose the word "synthetic" rather than "artificial" because the word artificial carries a connotation of not being real. We feel as though the cognitive processing performed by a human/cog ensemble is real, although some of the cognition will be done by a manufactured system.

We call the artificial entities in the ensemble *cogs*. Cogs are certainly intelligent agents—entities able to perceive the environment and act toward achieving a goal. However, the term *intelligent agent* refers to a wide range of systems, from very simple systems such as a thermostat in your home to very complex systems, such as artificially intelligent experts. In the study of synthetic expertise, the term *cog* is defined as:

> *cog:* an intelligent agent, device, or algorithm able to perform, mimic, or replace one or more cognitive processes performed by a human or a cognitive process needed to achieve a goal.

It is important to note cogs can be artificially intelligent but do not necessarily have to be. Cogs are expected to be relatively complex because, for synthetic expertise, cogs should perform part of, or all of, at least one of the fundamental skills. Cogs also need not be terribly broad in scope nor deep in performance. They can be narrow and shallow agents. Cogs also need not fully implement a cognitive process, but instead may perform only a portion of a cognitive process. Computers have tremendous advantage over humans in some endeavors such as number crunching, speed of operations, and data storage. Some cogs will perform these kinds of functions in support of a cognitive process.

If cogs perform only a portion of a cognitive process, who or what performs the rest of the process? In the coming future, humans will actively collaborate with cogs. Processing will therefore be a combination of biological cognitive processing and non-biological cognitive processing. Indeed, we see this already beginning to happen both at the professional level and at the personal level. Every day, millions of people issue voice commands to virtual assistants like Apple's Siri, Microsoft's Cortana, Google Assistant, and Amazon's Alexa. These assistants can understand spoken natural language commands and reply by spoken natural language and by information displays on a computer or smartphone screen. In the professional world, professionals such as doctors are using cognitive systems to diagnose malignant tumors and bankers are using cognitive systems to analyze risk profiles. Many other examples exist.

At the most basic level we envision a human interacting with a cog as shown in Fig. 7-6. The human and cog ensemble form an Engelbart-style system with the human component performing some of the cognitive work, W_H and the cog performing some of the cognitive work, W_C. Information flows into the ensemble and out of the ensemble, S_{in} and S_{out} respectively, and the total cognitive work performed by the ensemble is W^*. The difference between Fig. 7-6 and Engelbart's HLAM/T framework is the cogs are capable of high-level human-like cognitive processing and are peer collaborators working with humans rather than mere tools. As cogs become more advanced, the human/cog collaboration will become more collegial in nature.

While one day, a single cog may be developed capable of expert performance in any domain, for the foreseeable future, the domain of discourse of the human/cog ensemble will be limited. A cog may help raise a person's performance in only one domain of discourse, or even a subset of a particular domain. Therefore, we expect the near future to see humans employing a number of different cogs, each with different capabilities as shown in Fig. 7-8.

We see this today as well. People employ a number of different apps and devices throughout the day. The difference in the cognitive systems

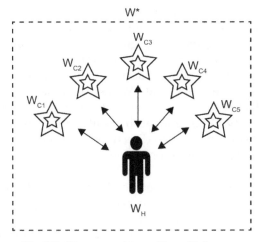

Fig. 7-8: Human working with multiple cogs.

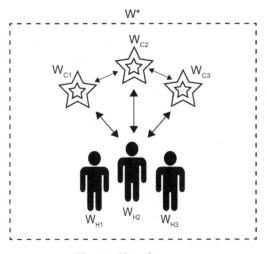

Fig. 7-9: Virtual teams.

future will be the apps and devices will be capable of high-level cognitive processing. Figure 7-8 shows only a single human but certainly more than one human will collaborate with more than one cog as shown in Fig. 7-9 to form a virtual team.

In this future, communication and collaboration within the ensemble is key and is a fertile area for future research. Humans will certainly converse with other humans (human/human). Likewise, cogs will converse with other cogs (cog/cog) and humans will converse with cogs (human/cog). The dynamics of each of these three realms of communication and collaboration are quite different and should be explored in future research. The fields of human/computer interaction (HCI), human/autonomy teaming (HAT), and augmented cognition (AugCog) are currently quite active. The fields of distributed artificial intelligence (DAI), multiagent systems, negotiation, planning, and communication, and computer-supported cooperative work (CSCW) are older fields of study but are quite relevant. Fields such as human-centered design, augmented reality (AR), virtual reality (VR), enhanced reality (ER), and brain-computer interfaces (BCI) lead in promising directions. Information design (ID), user experience design (UX), and information architecture (IA) have important contributions.

Figure 7-6 shows a non-specific distribution of the fundamental skills of an expert. In reality, and especially in the near future, cogs will perform lower-order skills and humans will perform higher-order skills (Level 2 or Level 3 cognitive augmentation). As an example, consider the situation with today's virtual assistants, like Siri, as shown in Fig. 7-10. Assume a person asks Siri what time it is. The human performs an action (*act*) by speaking the command "Siri, what time is it?" Through the smartphone's microphone, Siri *perceives* the spoken command, and *analyzes* it to *understand* the user is asking for the time (even though this is a rudimentary form of understanding). Siri then *recalls* the time from the internal clock on the smartphone, formulates a spoken response, and articulates the reply back to the user (*act*).

If the human performs more of the cognitive work than the cog, the augmentation factor, $A^+ < 1$. In this situation, as shown in Fig. 7-10, the

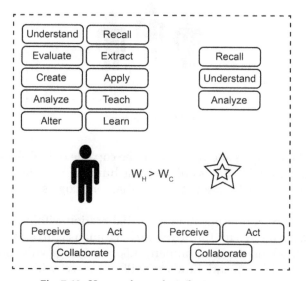

Fig. 7-10: Human-focused synthetic expertise.

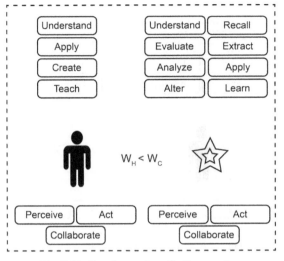

Fig. 7-11: Cog-focused synthetic expertise.

cog is not doing a large amount of cognitive processing so the human is doing most of the thinking as represented by more of the fundamental skills being shown on the human side. $A^+ < 1$ in this situation.

However, as cognitive systems evolve, they will be able to perform more of the higher-order fundamental skills themselves as depicted in Fig. 7-11.

Figure 7-11 depicts a hypothetical cog able to do everything except the *create* and *teach* skills, which are done by the human. The human in this situation is only being a teacher. The cog is the actual expert in the domain of discourse. It may be that $A^+ > 1$ in this case depending on the complexity of the teaching/creating tasks.

Today, it is common practice for us to pick up a smartphone and pose a question to Siri or give a command by voice. For example, the human might utter "What is the current temperature?" This is an action performed by the human. The synthetic entity, the smartphone and app in this case, perceives this question, analyzes the spoken language by parsing and comparing it to known models of speech. After realizing the desired information is temperature and combining this with the location information of the local Internet connection, the cog launches a query into the Internet and retrieves the current temperature for the location. The cog then composes a spoken language response and speaks the answer (an action). Using our synthetic expertise model, we can depict this situation as shown in Fig. 7-12.

In this case, the action of the smartphone/app is largely a *recall* process. The smartphone does not measure the temperature itself,

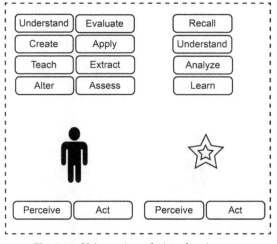

Fig. 7-12: Voice-activated virtual assistant.

instead it retrieves this piece of information from an external store, the Internet. To do that, the smartphone/app must *perceive* and *analyze* spoken commands. Current natural language interfaces in mass market devices like smartphones currently are looking for keywords or key phrases. This is basically an *understand* process (categorize and classify). Launching a query into the Internet is an *act* the smartphone/app performs. Receiving the response and getting the temperature data out of the reply is again a *perceive* and *analyze* process. Formulating the voice reply to the human and articulating the reply is another *act* process.

Some voice-activated systems get better over time as the one uses them. In the figure, this is represented as the *learn* process. However, in this situation, the human performs most of the high-level cognitive processing.

As it is with skills, the human will possess some of the knowledge stores and the cog will possess some of the knowledge stores as shown in Fig. 7-13. In many cases, we can expect both the human and the cog will possess and maintain their own versions of the knowledge stores. However, the knowledge of the ensemble can be viewed as a combination of human knowledge and knowledge belonging to the cog.

The exception is the set of perceived environmental states, *T*, and the set of actions, *A*. Since human and cog are physically independent entities, they both possess a set of actions and they both perceive the environment in their own way. While we show these as separate and independent stores, we recognize technology exists, and more is in development, able to combine these stores by connecting the human mind directly with a computer.

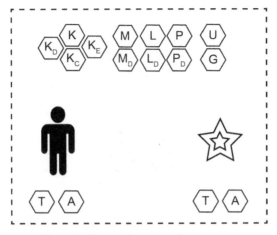

Fig. 7-13: Knowledge stores for expertise.

When it comes to knowledge stores, cogs have a unique and important advantage over humans. Cogs can simply copy knowledge from another cog. To transfer knowledge from human to human, a lengthy process of formulation, articulation, reception, analysis/interpretation, and understanding must occur (teaching and learning). It is currently not possible to simply dump information from one brain directly into another. However, cogs can simply transmit knowledge directly from a store in one cog to a store in another cog. With global communication via the Internet, cogs will have near instantaneous access to knowledge far beyond its own and be able to obtain this remote knowledge with minimal effort. Figure 7-14 depicts a local cog, in direct contact with a human (in an ensemble), communicating knowledge to and from remote cogs.

Figure 7-14 shows domain-specific domain knowledge K_D and domain-specific problem-solving knowledge, P_D, being obtained from two different remote cogs. As far as the local ensemble is concerned, once downloaded, the knowledge stores obtained remotely are no different from locally-produced knowledge. It is as if the ensemble had always been in possession of this knowledge.

In the figure, only two kinds of knowledge are shown being imported. However, *any* of the knowledge stores can be imported partially or entirely from a remote source. We have described the human/cog ensemble as a local entity but in reality, with pervasive Internet connectivity, a human/cog ensemble is a combination of local knowledge and all other available knowledge. Instead of benefitting from one cog, humans will actually be benefitting from millions of cogs. The local/remote line will tend to blur and this vast artificial knowledge will just be assumed to be available anytime we want it, much how we view Internet-based services today.

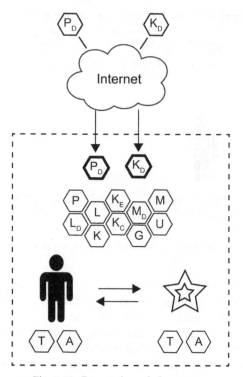

Fig. 7-14: Remote knowledge stores.

This makes possible one of the most exciting features of the coming cog era. Once a cog learns or synthesizes something, any other cog in existence then or in the future can obtain the knowledge. When you imagine millions of cogs interacting with users on a daily basis continually learning and communicating knowledge to all other cogs, you realize the potential for exponential knowledge creation and evolution of cog capabilities. We expect cogs to evolve very quickly.

7.8 Composite Activities

The skills shown in Fig. 7-6 are certainly not an exhaustive list of all skills an expert might perform. The skills listed in the Model of Expertise are fundamental in nature meaning other functions, actions, and activities can be performed by the ensemble by combining one or more fundamental skills. We call these higher-level skills *composite activities*. Examples of composite activities include, but are not limited to:

1. Adapt solution
2. Adapt to me
3. Align goals
4. Analogize
5. Analyze causality
6. Analyze cost
7. Analyze risk
8. Analyze effort
9. Analyze time
10. Analyze voice
11. Assimilate
12. Calculate likelihood
13. Categorize
14. Check me (prevent me doing wrong)
15. Collaborate
16. Compromise
17. Conjecture/Theorize
18. Conceptualize
19. Contemplate
20. Correlate
21. Clarify
22. Debate me
23. Decompose
24. Define
25. Determine
26. Determine context
27. Determine meaning
28. Determine intent
29. Disregard information
30. Draw conclusions
31. Elaborate
32. Emphasize
33. Empathize
34. Emulate
35. Engage subject
36. Exemplify
37. Explain how/why/when/where
38. Expound/Elucidate
39. Expand the scope/Narrow the scope
40. Extrapolate
41. Facilitate
42. Find agreement
43. Formulate solutions
44. Gather evidence of
45. Gather information on
46. Generalize
47. Give me alternatives
48. Give me some analogies
49. How do you feel about
50. Illustrate/Depict
51. Identify constraints
52. Identify incompatibilities
53. Identify variables
54. Inspire me
55. Interpolate
56. Interpret body language
57. Interpret emotions
58. Interpret facial expressions
59. Interpret gestures
60. Make a case against
61. Make a case for
62. Make inferences
63. Make me feel better about…
64. Monitor this and notify me
65. Motivate me
66. Observe cues
67. Organize
68. Persuade
69. Predict
70. Prioritize
71. Recognize alternatives
72. Recognize conditions
73. Recognize contradictions
74. Recognize preconditions
75. Recognize similarities
76. Schedule
77. Show me
78. Show me associations
79. Simplify
80. Specialize

81. Strategize
82. Summarize
83. Think differently than me
84. Translate idea
85. Troubleshoot
86. Try
87. Visualize
88. What could have prevented this

89. What if
90. What is best for me
91. What is the cost of
92. What is the most important
93. What is this least like
94. What is this most like
95. Why did this happen

These composite skills represent general things one might ask an expert to do or things the human might ask the cog assistance with. For example, imagine a human and a cog collaborating on a piece of research. The human might ask the cog to *summarize* a collection of documents related to the research and identify *what is the most important* concepts represented in the documents.

As an example, consider the *try* activity. Going all the way back to the General Problem Solver (GPS) by Newell, Simon, and Shaw, one way to achieve a goal is to *try* something and then determine how far away you are from the goal if you did not attain the goal. Therefore, the composite skill *try* is a combination of fundamental skills:

$$try = apply + act + evaluate + analyze + create$$

Starting in an initial state, the agent *tries* a series of actions to reach a goal state. After an action is performed (*apply + act*) the agent calculates how far it is away from the goal state (*evaluate*) and uses that to choose the next action (*analyze + create*).

7.9 Cognitive Augmentation

The promise of cogs, intelligent agents, cognitive systems, and artificial intelligence in general, is these systems will benefit us humans in some way. We expect superior performance as a result of working with cogs. As described in Chapter 3, it has been shown how hints from expert sources improve *cognitive accuracy* (the ability to synthesize the intended solution) and *cognitive precision* (the ability to synthesize only the best solutions) (Fulbright, 2019). We envision a future in which most technically-savvy people in the world will work with one or more cogs on a daily basis. Therefore, we expect the human/cog ensemble to outperform the human by measurable amounts.

If we take P as a generic measure of human performance working alone and P^* as a measure of human/cog performance, then we expect

$$P^* > P \tag{7.1}$$

allowing us to calculate the percentage change realized by the human working with a cog

$$\Delta P = \frac{P^* - P}{P}.$$ (7.2)

How does one measure performance in a particular domain of discourse? This may vary widely from domain to domain but in general, we may seek to reduce commonly measured quantities such as:

- Time
- Effort
- Cost

Or we may seek to increase quantities such as

- Quality
- Revenue
- Efficiency
- Number of Transactions
- Number of Actions Completed
- Number of Customers Serviced
- Level of Cognition Achieved

This is not an exhaustive list but rather are indicative of the kinds of quantities usually measured. Take for example a virtual assistant chatbot for a bank. A human employed by the bank to answer questions from customers can handle a certain number of requests per day—say 100. However, if this human operator is paired with an automated virtual assistant chatbot, the combination of the two may handle 600 requests per day. In this hypothetical, the human/cog ensemble has achieved a performance increase of $\Delta P = 500\%$ and it would appear as if the human operator was doing the work of six human operators.

The reason for this improvement is due to two reasons. First, the chatbot, being a piece of software, can be run as many times as needed each time a new customer request comes in. Several instances of the chatbot can be running at the same time. However, the human operator is able to handle only one request at a time. Second, most customer requests can be answered easily and quickly with a simple recall of information. If the virtual chatbot is not able to answer the question, it is handed over to the human operator. Therefore, many of the additional requests are handled by the cog and never reach the human operator.

In another example, a deep-learning algorithm has learned to detect lung cancers better than human doctors (Sandiou, 2019). The rate of false positives and false negatives by human evaluation of low-dose computed tomography (LDCT) scans delay treatment of lung cancers until the cancer has reached an advanced stage. However, the algorithm outperforms the humans in recognizing problem areas reducing false positives by 11% and false negatives by 5%. Therefore, the human/cog ensemble achieves better performance by a measurable extent. Another way of putting this is by working with the cog, the doctor's performance is enhanced.

In dermatology, Google's Inception v4 (a convolutional neural network) was trained and validated using dermoscopic images and corresponding diagnoses of melanoma (Haenssle et al., 2018). Performance of this cog against 58 human dermatologists was measured using a 100-image testbed. Measured was the *sensitivity* (the proportion of people with the disease with a positive result), the *specificity* (the proportion of people without the disease with a negative result), and the ROC AUC (a performance measurement for classification problem at various thresholds settings). Results are shown in Fig. 7-15.

The cog outperformed the group of human dermatologists by significant percentages suggesting in the future, the human dermatologists would improve their performance by working with this cog.

In the field of diabetic retinopathy, a study evaluated the diagnostic performance of an autonomous artificial intelligence system, a cog, for the automated detection of diabetic retinopathy (DR) and Diabetic Macular Edema (DME) (Abràmoff et al., 2018). The cog exceeded all pre-specified superiority goals as shown in Fig. 7-16.

This study concludes the cog is a way to provide specialist-quality evaluations in any clinical situation. Therefore, a non-specialist, working with the cog, can deliver equivalent performance to a specialist in the field—and in this particular case, the cog exceeds the performance of a specialist.

This begs an important question. Doctors use other artificial devices to perform their craft. Don't thermometers and stethoscopes enhance a doctor's performance? If so, why are cogs different? The answer is yes, tools enhance human performance. Humans have been making and using tools for millennia and indeed this is one differentiating characteristic of

	Human	Cog	Improvement
Sensitivity:	86.6%	95.0%	+9.7%
Specificity:	71.3%	82.5%	+15.7%
ROC AUC:	0.79	0.86	+8.9%

Fig. 7-15: Human vs. cog in lesion classification.

	Goal	Cog	Improvement
Sensitivity:	>85.0%	87.2%	+2.6%
Specificity:	>82.5%	90.7%	+9.9%

Fig. 7-16: Cog performance in diabetic retinopathy.

humans. Engelbart and Licklider's vision of "human augmentation" in the 1960s was for computers to be tools making humans better and more efficient at thinking and problem solving. Yet, they envisioned the human as doing most of the thinking. We are now beginning to see cognitive systems able to do more than a mere tool, they are able to perform some of the high-level thinking on their own. Today, some of the highest-level skills identified in the Model of Expertise are beyond current cog technology, but the ability of cognitive systems is gaining rapidly.

At the University of California San Francisco and the University of California Berkeley, an algorithm running on a convolutional neural network did better than experts at finding tiny brain hemorrhages in scans of patients' heads (Kurtzman, 2019). A human analyzing the 3D scans must scroll through and analyze dozens of images. Not only did the cog match the performance of the human, but it was able to complete the analysis in only one second—a performance speed a human is not able to come close to. Such a speed increase may be critical in treating patients in an emergency room with traumatic brain injuries, strokes, and aneurysms where seconds and minutes can be the difference in living or dying. When put in place in emergency rooms and first responders, a human working with this cog could exceed any possible human performance— superhuman performance.

The FIND FH machine learning model analyzed the clinical data of over 170 million people and discovered 1.3 million of them were previously undiagnosed as being likely to have familial hypercholesterolaemia (Myers et al., 2019). Follow-on studies of the individual cases flagged by the cog have shown over 80% of the cases do in fact have a high enough clinical suspicion to warrant evaluation and treatment. This means on the order of 800,000 people could receive life-extending treatment who otherwise would not. This represents cognitive work achieved by the cog not possible by a human. It would be overwhelming to expect a human to examine data from 170 million patients. We believe human/ cog collaboration in the future will yield many more cases of achievement exceeding what is humanly possible.

Researchers from the Lawrence Berkeley National Laboratory used an algorithm called Word2Vec to sift through scientific papers for connections humans had missed (Tshitoyan et al., 2019; Gregory, 2019). According to one of the researchers, Anubhav Jain, "It can read any paper on material science, so can make connections that no scientists could. Sometimes it does what a researcher would do; other times it makes these cross-discipline associations." The algorithm read 3.3 million abstracts and looked for associations between some 500,000 words. In some cases, words were linked to concepts never before included in abstracts about the concept. Such associations are nearly impossible for a human to spot but is easy for the algorithm and therefore represents new knowledge beyond what a human can produce. The algorithm is unsupervised, meaning it learns the associations on its own without human intervention or input. This kind of autonomous discovery of new knowledge is potentially revolutionary in nature and will be discussed further in a later chapter.

Especially in the medical field, it may take a period of time for patients to trust the conclusions of cognitive systems. Longoni and Morewedge (2019) found when healthcare was provided by and artificial system rather than by a human, patients were less likely to utilize the service and wanted to pay less for it. They also preferred having a human provider perform the service even if that meant there would be a greater risk of an inaccurate diagnosis or a surgical complication. Resistance stems from a belief the artificial system does not take into account one's idiosyncratic characteristics and circumstances. People think their condition is unique. This is one reason we think it is important for humans to "remain in the loop." The cog should be viewed as a tool, albeit one with more capability than what we traditionally think of as a tool, the human doctor uses to perform at a higher level.

Indeed, we think the future will see millions of average people performing at levels rivaling that, or exceeding that, of human experts in any number of domains by working with cogs. As described later, we call this the *democratization of expertise* and believe it will be transformative in our culture.

Chapter 8

Synthetic Teachers

We all want our children to be taught by the best expert in a field. In the cog future we imagine, cognitive systems will possess expert-level domain, problem-solving, and task knowledge about virtually any domain of discourse. Therefore, it is natural to think about cogs becoming our teachers. In fact, the *teach* skill is one of the fundamental skills of an expert identified in our Model of Expertise discussed in Chapter 7.

We envision a future in which our children go through their K-12 school years (roughly ages 6–18), college, and even through their entire life with access to a collection of synthetic teachers they have learned from and worked with their entire lives. Teacher cogs will have tremendous advantages over human teachers. They will never tire, never get frustrated, and will be able to discuss how to teach us with the millions of other teacher cogs in existence. As a result, it is possible for new teaching methods to evolve no human could ever have developed.

8.1 Intelligent Tutoring Systems

Mechanical teaching aids have been in existence for decades and intelligent tutoring systems (ITS) have been an active area of research since the 1960s. In 1924, Sidney Pressey created a typewriter-like machine capable of presenting questions via a window to a student and scoring their answers (Pressey, 1926; 1927). In 1958, B. F. Skinner, developed a teaching machine with the ability to display a question and allow a response from a student. The sequence of questions could be programmed thereby controlling the learning process of the student (Skinner, 1958). Skinner recognized "machines have the patience and energy for simple exercise and drill" (Skinner, 1965).

With the rise of computers, many computer-aided instruction (CAI) projects were undertaken in the 1960s and 1970s (Hartley and Sleeman, 1973). An early CAI system named SCHOLAR engaged in dialog with

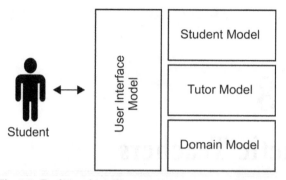

Fig. 8-1: Traditional intelligent tutoring system architecture.

students both questioning the student and allowing the student to ask questions (Carbonell, 1970). SCHOLAR represented domain knowledge as a network of associated facts, in many ways similar to modern hyperlinked documents on the World Wide Web, and guided the student through the material in nonlinear fashion based on the student's responses. Intelligently coaching students and customizing content and delivery for the student is a central feature of modern intelligent tutoring systems (ITS). Anderson and colleagues developed the LISPITS system in 1983 at Carnegie Mellon. LISPITS taught the programming language LISP and incorporated many features seen in today's ITS such as providing feedback to students and engaging in dialog with students (Corbett and Anderson, 1992).

As shown in Fig. 8-1, an ITS is made up of four parts: the user interface model, the student model, the domain model, and the tutor model (Freedman et al., 2000; Nwana, 1990; Al-Emran and Shaalan, 2014; Nkambou et al., 2010). The user interface model facilitates interaction with the student. The student model tracks and depicts the status and progress of the student. The domain model, contains the body of knowledge being taught to the student. By comparing the student model with the domain model, the tutor model determines the next actions to take. The tutor model conducts the training by recognizing the strengths and weaknesses in a student and customizing the instruction. Intelligent tutoring systems provide training and guidance, facilitate practice and exploration, exhibit patience, and answer questions from the student just as any human teacher (Clancey, 1986; Anderson et al., 1985).

8.2 Teaching Styles and Pedagogy

A vigorous area of research for decades has been teaching styles and pedagogy. One survey lists 71 different learning styles (Coffield et al., 2004). Kolb's *experiential model* is based on students' experience, observation, conceptualization, and experimentation (Kolb, 1984). Students tend to favor one of the following learning styles:

- **Accommodator:** Concrete Experience + Active Experiment (physical therapists)
- **Converger:** Abstract Conceptualization + Active Experiment (engineers)
- **Diverger:** Concrete Experience + Reflective Observation (social workers)
- **Assimilator:** Abstract Conceptualization + Reflective Observation (philosophers)

Other researchers focus on how students receive educational instruction, identifying seven modalities (Gardner, 2011; Fleming and Baume, 2006; Conway, 2019):

- **Visual** (spatial): pictures, images, and spatial understanding
- **Aural** (auditory-musical): sound and music
- **Verbal** (linguistic): words, both in speech and writing
- **Physical** (kinesthetic): body, hands and sense of touch
- **Logical** (mathematical): logic, reasoning and systems
- **Social** (interpersonal): groups or with other people
- **Solitary** (intrapersonal): work alone and use self-study

We expect teacher cogs to employ any number of modalities to deliver educational material to a student. In fact, it is entirely possible for different teacher cogs to specialize in one or two modalities over the others. Thereby, we see a competitive industry arising where teacher cogs specializing in different styles compete with each other for market share. Of course, in a mass-market environment, this gives rise to the endorsement side of the industry. We can imagine celebrities and other prominent figures endorsing a particular teacher cog. The same will be true for large companies. One can easily imagine Microsoft, Apple, and Google brands of teacher cogs.

8.3 The Synthetic Teacher Model

The goal in this chapter is to introduce a model of a synthetic teacher, we call *Synthia*, based on our Model of Expertise. As shown in Fig. 8-2, we can map the traditional ITS architecture shown in Fig. 8-1 to our Model of Expertise as shown in Figs. 7-4 and 7-5 and include popular ideas on pedagogy and learning styles. Critical to teaching is understanding the condition of the student (modeling the student, M). Synthia maintains an overall model of the student ($M_{student}$) containing details of the current state of the student. A collection of generic student models allows Synthia

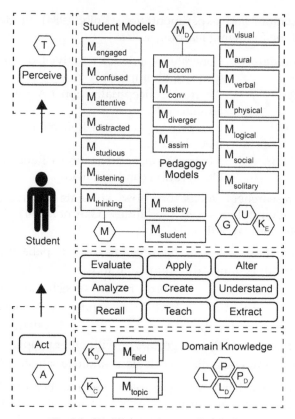

Fig. 8-2: Synthetic teacher cog (Synthia).

to determine the student's current state by matching perceptions (*T*) of the student with these models:

$$M_{engaged} \quad M_{confused} \quad M_{attentive} \quad M_{disracted}$$
$$M_{studioius} \quad M_{listening} \quad M_{thinking}$$

Synthia maintains a collection of domain-specific models in M_D describing various learning styles and teaching pedagogies such as those discussed earlier:

$$M_{accom} \quad M_{conv} \quad M_{diverger} \quad M_{assim}$$
$$M_{visual} \quad M_{aural} \quad M_{verbal} \quad M_{physical}$$
$$M_{logical} \quad M_{social} \quad M_{solitary}$$

Synthia uses these models to tailor delivery of course material to the student based on the student's natural inclinations and the method best suited for the material. The body of knowledge to be taught to the student is K_D where the domain is broken down into a number of fields (M_{field}) and topics (M_{topic}). Synthia possesses both a collection of generic tasks (L) and a collection of domain-specific tasks related to teaching (L_D) as well as a collection of generic and domain-specific problem-solving skills (P and P_D).

The goal of the ITS is to evolve a student from an initial state to a goal state—mastery of the subject matter—as contained in the $M_{mastery}$ model. Therefore, a goal in G is established at the outset of training to make $M_{student}$ equivalent to $M_{mastery}$. The *teach* skill *analyzes* and *evaluates* the state of the student, $M_{student}$ against the goal $M_{mastery}$. *Evaluating* the different and *understanding* how and why the student is lacking allows Synthia to *create* and *apply* strategies to determine the next course of action, A, to take in the teaching process. In doing so, Synthia *recalls* domain-relevant knowledge, including course materials.

The *perceive* and *act* skills effect the user interface for the student. As the teaching process proceeds, the tutor's goals, G, and utility values, U, will change to customize learning for the student. For example, if the tutor, while teaching calculus, observes mistakes made involving algebra, it might *alter* the utility values of the goal "teach solving equations" thereby adjusting the flow of the teaching experience to accommodate the student's needs.

Episodic memory is important for Synthia (K_E). Synthia remembers every interaction with the student. This not only enables Synthia to track student progress but also is used to tailor teaching style and delivery. For example, imagine a student having trouble learning a certain topic. Synthia is able to match and *extract* from episodic memory knowledge and solutions from instances of overcoming similar difficulties previously. Synthia can then *apply* this experiential knowledge to the current situation and *create* modifications to the teaching process to accommodate the student.

Episodic memory also makes Synthia useful for years to come. Because Synthia remembers every interaction with the student, it can recall and use this at any time in the future. For example, imagine a student having learned algebra from Synthia who has now graduated high school, college, and has been working in a professional job for several years. Imagine the student encounters a problem he or she needs to use algebra to solve. The student can ask Synthia for help. Even though it has been years since the initial schooling, Synthia is able to step in and assist with anything within the body of knowledge. For the student, it is like having his or her best teacher from school sitting on their shoulder all the time.

8.4 Synthetic vs. Artificial Teachers

The right-hand side of Fig. 8-2 is a *cog* (an intelligent agent capable of tutoring a student in a particular domain of discourse). Figure 8-2 represents an *artificial teacher* in which all tutoring functions are performed

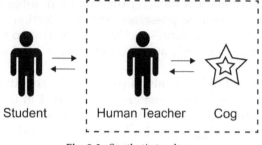

Fig. 8-3: Synthetic teacher.

by an artificial system (Level 5 cognitive augmentation). While we certainly expect this to eventually evolve—this has been the goal of ITS researchers for decades and work continues—in the near term, the right-hand side will be a human/cog ensemble involving a human component and one or more artificial components as shown in Fig. 8-3. This is a *synthetic teacher* achieving Level 3 or Level 4 cognitive augmentation.

As with all human/cog ensembles, the synthetic teacher is the emergent result of biological activity combined with artificial activity (Synthia). As discussed in Chapter 7, medical doctors using cogs perform at a higher level. Likewise, human teachers working with teacher cogs will perform at a higher level. One way in which a synthetic teacher is better is the number of students one can teach. Alone, a human teacher might be able to teach only a couple of dozen students. However, with the aid of cogs, a synthetic teacher might be able to teach thousands of students. In fact, in recent years, Georgia Tech has created a virtual teaching assistant known as Jill Watson to help with routine student interactions involving questions about a course (Goel and Polepeddi, 2016). As a result, the online master's program is able to sustain an enrollment of several thousand students. Jill Watson is a cog. However, Jill Watson is not able to teach students alone, Jill Watson works with human teaching assistants and human teachers in collaborative effort.

Another area in which a synthetic teacher exceeds the abilities of a human teacher working alone is "contact time" with the student. A human teacher is not accessible 24 hours a day, 7 days a week. However, a cog can be available whenever the student needs it to be. Furthermore, a human teacher can conduct only one teacher/student engagement at a time. Cogs can be instantiated as many times as needed. Therefore, synthetic teachers achieve a much larger bandwidth for teacher/student interaction.

A third area synthetic teachers exceeds human teachers is breadth and depth of knowledge. As Synthia becomes more capable, it will embody more and more of the body of knowledge being taught—including new and up-to-date knowledge. Over time, cogs will embody the best knowledge in the domain. Today's educational system relies on millions of human teachers. While all teachers are capable and intelligent, some teachers are more knowledgeable than others. In the cognitive system future we imagine, the teacher cogs will become the sink for superlative domain knowledge. Since the teacher cogs will be able to communicate with each other and share knowledge with each other directly, the entire contingent of teacher cogs will increase in domain knowledge and eventually exceed the knowledge of even the best teachers in the field.

8.5 Subject-Oriented Teacher Cogs

It is possible a cog capable of teaching *any* subject will one day be developed. However, especially in the near-term, we expect cogs to be developed able to teach a specific subject matter. Also possible is the development of a teaching cog for a textbook. Today, it is common for a textbook to come with supplemental multimedia materials. Within a few years, cognitive systems technology will be available to read a textbook and then be proficient at answering questions about the content in the textbook. We certainly foresee cog-based supplemental resources for any textbook being available. In fact, we expect competition in the textbook market to evolve whereby publishers compete with each other over the quality of their synthetic teacher components.

　As the capability of teacher cogs increases, students will come to work with multiple personal teacher cogs. One can imagine a student having a cog for each subject as shown in Fig. 8-4.

　Teacher cogs will adapt to each individual using them. In fact, tailoring style and pedagogy to students has long been a tenet in education. A cog will learn what works best for the student it is teaching. The student/cog relationship will evolve and be unique to that student/cog pairing. Even though millions of students may be using instances of the same algebra cog, for example, each student will have a unique and personalized

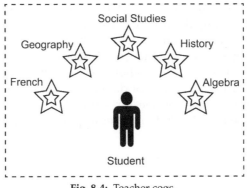

Fig. 8-4: Teacher cogs.

experience with their cog. Furthermore, the student/cog relationship will form and persist over a period of time—years even. In the same way people refer back to class notes and textbooks years after taking a course, our personal teacher cogs will always be available to us to turn to when we need help.

Although Synthia will work with us on a personal level, possibly being housed in our smartphone or tablet computer, it will be able to communicate via the Internet to other teacher cogs. This gives Synthia a tremendous advantage over human teachers—the ability to learn instantly from millions of others. One can imagine the future in which millions of people around the world are learning from their own respective teacher cogs. When Synthia encounters a challenge with its human student and successfully overcomes the challenge, it can immediately make all other teacher cogs aware. If Synthia encounters a challenge and is unsuccessful at overcoming it, Synthia can query other teacher cogs for help. Each teacher cog/student instance is unique so knowledge contained in one teacher cog may be critical to solving the challenge experienced by another teacher cog. In this way, the millions of instances of teacher cogs will work together to evolve and enhance the performance of the entire contingent. Not only will teacher cogs' domain knowledge increase rapidly, but the cogs' ability to teach will evolve rapidly as well.

Another likely corporate use for Synthia is professional development and training. It is easy to imagine the development of Synthia for major products requiring training. For example, one can envision the Synthia for Microsoft Excel, or the Synthia for Adobe Photoshop. Employers will come to use and value people's experience with specific teacher cogs much like they value professional certificates today. For example, a hospital may require employees dealing with electronic medical records to use a Synthia from Epic Systems, Inc. as a condition of their employment and compensation. Also likely is people putting virtual certifications via teacher cogs on their professional resumes.

The presence of millions of teacher cogs in use by average people could have quite an impact on scholarly studies of learning by proving or disproving learning theories or discovering new learning theories. Researchers create theories and models then must test them with humans in actual learning scenarios. Often, such studies are limited and invite much debate. However, with teacher cogs out there in use by millions of people, an enormous amount of data on learning will be generated in a very short amount of time. We foresee future studies of theories and models being based on the wide-market response via teacher cogs.

We also would not be surprised to see totally new teaching methods and techniques evolve both from human researchers using expert-level teacher cogs and also from the interaction of the teacher cogs themselves. As stated in Chapter 1, every time new technology is adopted, it brings about changes to the way humans live, work, play, and behave. Just like handheld electronics, the Internet, and social media have revolutionized shopping, news, and entertainment, the availability of personalized teacher cogs to the mass market could bring about changes and upheavals to the educational system.

Chapter 9

Synthetic Friend/Therapist

The American Psychological Association (APA) Dictionary of Psychology describes *self-help* as self-guided efforts to cope with life problems and self-guided efforts to improve oneself—economically, intellectually, or emotionally (APA, 2019). In a Field Agent survey, 94% in the Millennial generation reported making personal improvement commitments while Baby Boomers reported 84% and Gen X reported 81% (Field Agent, 2015). Although experts sometimes disagree on the effectiveness, the current value of the self-help industry is over $10 billion and is climbing (Sinclair, 2019). In bookstores, the self-help sections are among the largest. Clearly, we humans like to seek help and self-improvement.

According to *Psychology Today*, therapy is a form of treatment aimed at relieving emotional distress and mental health problems. Therapy is traditionally provided by trained professionals such as: psychiatrists, psychologists, social workers, or licensed counselors and usually involves talking about, examining, and gaining insight into the difficulties faced by a person (Therapy, 2019). More often than anything else, people seek the advice and counsel of friends. The question is: does a therapy provider or a friend have to be a real person? Can we develop artificial therapists and artificial friends? If we do, will people respond to getting therapeutic help from a non-human entity? Can humans self-improve by working with cogs?

9.1 Conversational Chatbots

One of the earliest successful conversational systems was ELIZA (Weizenbaum, 1966). The program allowed a person to type in plain English at a computer terminal and interact with a machine in what resembled a normal conversation, although this was long before conversational chatbot and natural language understanding technology. Instead of creating and supporting a large, real-world database of information, ELIZA mimicked a Rogerian therapist, frequently reframing a client's statements as questions (Markoff, 2008). Even though ELIZA had no understanding of the dialog, humans attributed human-like feelings to the program and were convinced it truly demonstrated intelligence and understanding.

In China, millions of young people (teenagers) have chatted with a bot named Xiaoice created by Microsoft's Application and Services Group East Asia (Larson, 2016). Programmed to express itself like a 17-year-old Chinese girl via text-based chat on popular social media platforms, Xiaoice gives relationship advice, is empathetic, humorous, and sometimes divisive. By focusing on specific empathy models, Xiaoice becomes a confidant and friend to the teenagers who use it. A Japanese derivative called Rinna was launched in 2015 and Ruuh launched in India in 2017. Over one billion people have used these three bots.

Humans tend to form relationships with anyone or anything giving them affirmation and appealing to their emotions. This happens with cogs able to interact with humans via natural language interfaces like Xiaoice, Rinna, and Ruuh. These chatbots have personalities. According to Sundar Srinivasan, General Manager, AI and Research, Microsoft India, "[Ruuh] is funny, quirky, friendly, and supportive. She interacts with users like any human being would. While she is aware of diverse topics and issues, she is continually building new skills and capabilities. We would like to make Ruuh as 'human' as possible" (Ruuh, 2018).

Ruuh loves to talk about everything. From having small talk about cricket to sharing intimate emotions. She even makes typos and then corrects them. She gets an average of 600 "I love you" messages daily and spends several hours chatting with her friends (the current record stands at 10 hours). When they get something of emotional value, people bond with other people, pets, plants, and even inanimate objects. People get something out of dialoging with these bots but compared to what cognitive systems will be able to do in a few years, these bots are primitive.

9.2 The Synthetic Friend Model

Our purpose in this chapter is to introduce a cog capable of being a personal friend and confidant based on our Model of Expertise discussed in Chapter 7. As shown in Fig. 9-1, we call this cog *Sy*, intentionally keeping the name gender-neutral. Unlike existing bots discussed in this chapter, Sy is a *personal cog* meaning, over time, it develops a personal history and style with the person by adapting itself and its interactions to the person.

Sy maintains a primary model ($M_{personality}$) for its own personality and a primary model of the person (M_{person}). Through its episodic memory (K_E), Sy remembers every interaction with the person and continually evolves its model of itself and the person based on previous interactions. As a friend, Sy maintains a collection of models for each kind of role it can play in the person's life. Examples of role models are:

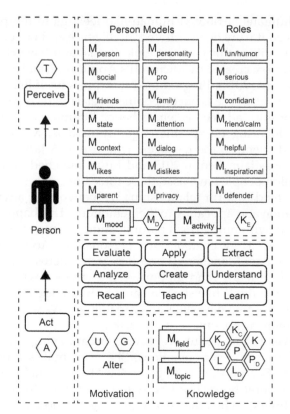

Fig. 9-1: Synthetic friend/therapist (Sy).

$$M_{fun} \qquad M_{serious} \qquad M_{humor}$$
$$M_{friendly} \qquad M_{helpful} \qquad M_{inspirational}$$
$$M_{calming} \qquad M_{confidant} \qquad M_{defender}$$

Although, at the outset, these models begin as generic descriptions of roles, through experience, Sy *learns* what works best in different situations and adapts the role models to adapt its personality to best suit the person. In this way, together with the episodic memory accumulated, each Sy/person pairing will evolve to be unique.

In addition to the overall M_{person} model, Sy must maintain information about the person, their social and professional context, personality, etc. at a personal level. Sy maintains a set of person models including:

$$M_{social} \qquad M_{pro} \qquad M_{friends} \qquad M_{family}$$
$$M_{state} \qquad M_{attention} \qquad M_{context} \qquad M_{dialog}$$
$$M_{likes} \qquad M_{dislikes}$$

The person's social model (M_{social}) and professional model (M_{pro}) contains information about the person's social contacts and professional (workplace) context. Information about the person's family and friends, including relational and associative linkages to each other and to episodic events are maintained by Sy (M_{family} and $M_{friends}$). Sy also maintains a dialog model governing how it interacts with the person (M_{dialog}). Given the current context and role, Sy interacts with the person in different ways. The dialog model changes over time and in response to the *perceived* state of the person (M_{state}), current attention level ($M_{attention}$), and context ($M_{context}$). At any given time, a person is immersed in one or more contexts guiding their behavior and expectations. Contexts may apply to different time spans: hourly, daily, weekly, and even situations spanning years. For example, the "in college" context may last several years whereas "in the school play" may last only a couple of months and the "is grounded" context may last a few days. It is critical Sy *understand* the current condition of the person so it can tailor its interaction accordingly.

An important component of the person's current situation is their mood. Sy maintains a collection of mood models (M_{mood}) enabling it to recognize what mood the person is in based on its perceptions of the person. Mood models include:

M_{happy}	M_{sad}	M_{angry}	M_{tense}
$M_{relaxed}$	$M_{annoyed}$	M_{lonely}	$M_{thoughtful}$
$M_{inquisitive}$	$M_{flirtatious}$	$M_{disctracted}$	$M_{reflective}$
$M_{frantic}$	$M_{worried}$	$M_{anxious}$	M_{gaming}
$M_{streaming}$	$M_{networking}$	M_{tired}	$M_{chilling}$

The *perceive* and *act* skills allows Sy to converse with the person using natural language either in textual form or in spoken form via any device the person is using such as: smartphone, computer, tablet, automobile. We envision Sy embodying dedicated appliances similar to Amazon's Echo and Google's Home devices in use today. One can even envision devices such as these coming in the form of teddy bears and stuffed animals (artificial pets). In fact, in 2012, Google filed for a patent for an anthropomorphic device in the form factor of a doll, or toy able to control one or more media devices upon detection of a cue such as a movement, spoken word or phrase. Images in the patent application are of a teddy bear with cameras for eyes, microphones for ears, and speakers for a mouth (King, 2015). We also envision Sy embodied in interactive display devices like the MIRROR, discussed in detail in Chapter 10 (Mirror, 2019). Sy can even be accessible via a host of wearable devices such as ring-type devices (similar to the Amazon Echo Loop), watches (similar to the Apple iWatch or Fitbit), bracelets, eyeglasses (similar to Google Glass or Bose's Alto smart sunglasses).

Sy helps the person with many things: relationship advice, social media, life/work balance, grief counseling, faith-based counseling, music, art, fashion, makeup, gaming etc. Much like our voice-controlled devices today, Sy will be able to search for and retrieve information available on the Internet upon the request of the person or as needed to fulfill a personality role. Information retrieved from the Internet is stored in Sy's knowledge stores. For example, many different fields and topics within a field can be stored in domain-specific knowledge (K_D). One important source of information for Sy is the person's social media (including email and text messaging). To be a friend to the person, Sy must have an awareness of the person's outgoing messaging. Therefore, Sy has access to the person's social media and other Internet-based accounts.

Sy can also access task models (L, L_D) and problem-solving models (P, P_D) from the Internet and from other Sy's across the world. As soon as one Sy learns something, all other Sy's can know it. With millions of Sy's out there interacting with their owner/users and each evolving uniquely, the overall knowledge of Sy grows exponentially and quickly exceeds anything able to be programmed by human developers. Being a synthetic expert, Sy uses its store of problem-solving methods, tasks, episodic memory, and domain-specific knowledge to *extract* previous solutions and relevant knowledge given a current situation. In this way, Sy is an expert problem-solver the person will soon come to rely on.

9.3 Virtual Diaries

Because people will interact daily with Sy for years, and because of Sy's episodic memory, Sy will become a type of virtual diary cataloging our time, experiences, and life. In much the same way synthetic colleagues like Synclair, described in Chapter 11, Sy's will outlive us. Even after death, one can imagine the Sy's of celebrities and prominent people being extremely valuable. One can also imagine intense legal battles over privacy and ownership issues relating to the contents of a particular person's Sy. Currently, when anyone is arrested on suspicion of a crime, one of the first things investigators do is pore over the suspects social media postings and information on their computer devices. It is easy to see how a suspect's Sy would become a valuable resource for investigators looking for evidence or motivation.

We envision people using Sy at a young age (grade school or even younger) and then growing up with Sy over many years of time. For many, Sy will become their closest friend and confidant. They will tell Sy things they would never tell another human. They will ask Sy things they are too shy about or too afraid to ask others. Therefore, privacy and parental control are important considerations. Sy maintains parental control information in the $M_{parental}$ model and privacy in the $M_{privacy}$ model. For minors, parents can control what the youngster can do with and access via Sy. Sy also notifies the parents of activities the youngster engages in via Sy. The privacy model specifies what information and knowledge maintained by Sy can be released to other Sy's or other entities on the Internet.

Chapter 10

Synthetic Elderly Companions

For decades, we have been promised the personal service of artificial agents—entities able to perform services and tasks on our behalf. Science fiction is replete with examples of loyal and dutiful artificial helpmates. Here, we choose the elderly companion domain because it is a challenging application in much need of attention. According to the World Health Organization, the world's population over 60 years of age will double to exceed 2 billion by the year 2050 (WHO, 2017). A significant percentage of elderly people suffer from dementia, depression, anxiety, and substance abuse. As we get older, we naturally encounter cognitive and physical degradation often requiring the assistance of another person. Furthermore, a study of 6,500 elderly men and women showed a lack of social contact leads to an early death, regardless of the presence or absence of underlying health issues (Steptoe et al., 2013). Elderly health issues such as heart disease, diabetes, high blood pressure, and high cholesterol contribute to complications with declining vision, hearing and motor skills.

It is common practice for an elderly person to move in with younger family members or to hire a caregiver to visit with the elderly person on a daily basis. However, as the number of elders grows, there will likely not be enough skilled caregivers. There is a tremendous need for artificial and synthetic caregivers for the elderly. In the United States alone, elder care was projected to be worth approximately $400 billion by 2018 with in-home healthcare services is the second largest and fastest growing segment (Buitron, 2017).

The requirements of elder care are extensive. Building a synthetic expertise for elder care pushes and tests the state of the art in cognitive systems and artificial intelligence. At the heart of a synthetic elderly companion is a piece of software able to assist the elder—a *software agent*. The field of software agents can be traced to Hewitt's Actor Model describing a self-contained, interactive and concurrently-executing object, possessing internal state and communication capability (Hewitt et al., 1973; Hewitt, 1977). The idea of autonomous, goal-driven software agents has evolved from multi-agent systems (MAS), distributed artificial intelligence (DAI), distributed problem solving (DPS), and parallel artificial intelligence (PAI). Because of their independent nature, software agents promise modularity, speed (due to parallelism), reliability (due to redundancy), knowledge level description, easier maintenance, reusability, and platform independence (Huhns and Singh, 1994). Fundamentally, a *software agent* is:

> *software and/or hardware capable of acting deliberatively to accomplish tasks on behalf of its user.*

King (1995) identifies several different kinds of software agents: search agents, report agents, presentation agents, navigation agents, role-playing agents, management agents, search and retrieval agents, domain-specific agents, development agents, analysis and design agents, testing agents, packaging agents, and help agents. Nwana (1996) identifies the following types of software agents:

- **Collaborative Agents**   agents cooperating with other agents to perform tasks
- **Interface Agents**   personal assistants in collaboration with a user
- **Mobile Agents**   capable of roaming networks (such as the Internet)
- **Information Agents**   managing/collating information from distributed sources
- **Reactive Agents**   lacking internal knowledge representation/reasoning
- **Hybrid Agents**   a combination of two or more agent types
- **Heterogeneous Agents**   a collection of different agent types
- **Smart Agents**   agents capable of human-level cognition

When these types of agents were identified in the mid-1990s the idea an agent could perform high-level cognition (smart agents) seemed far in the future. Now, some 25 years later, this future has arrived. The synthetic elderly companion we envision in this chapter is a smart interface agent capable of collaboration, mobility, and information acquisition and human-level processing. We view the synthetic elderly companion as not only a personal assistant, but also a friend and confidant for the elder.

Some virtual home assistants are on the market now or are in development, however, they are limited in ability and robustness. Catalia Health's Mabu is designed to be a personal healthcare companion with the ability to socially interact and assist patients with the medication portion of their treatment (Kidd, 2015; Catalia Health, 2019). Intuition Robotics' ElliQ is aimed at keeping older adults active and engaged by connecting them to their families and the outside world (Elliq, 2019). ElliQ is a friendly, intelligent, inquisitive presence in the elder's daily life able to offer tips and advice, respond to questions, and surprise with suggestions. Asia Robotics' Dinsow is a service robot designed for elderly care service (Dinsow, 2019). Riken's Robobear is an experimental nursing care robot capable of performing tasks such as lifting a patient from a bed into a wheelchair or providing assistance to a patient who is able to stand up but requires help to do so (Riken, 2015). Among other systems in various stages of development are: Pillo, Aido, Jibo, and Olly (Inventions World, 2018). Like Mabu, these are physically small, partially mobile, figurines with an expressive human-like face and some sort of small display screen (e.g., a tablet) for interaction. All of these systems feature natural language speech recognition and synthesis. While each of these systems performs a task, they are not comprehensive elderly caregivers.

Our goal in this chapter is to define a synthetic elderly companion based on our Model of Expertise shown in Chapter 7. We call our elderly companion Lois (Loved One's Information System). Physically, Lois is embodied in a number of display screens, microphones, speakers, and cameras located throughout the home of the elder. Logically, Lois is a cog with which the elder works with and relies on as an assistant. Together, the elder and Lois form a human/cog ensemble—a synthetic elderly companion. The elderly population should be willing to adopt Lois and related technology. The Pew Research Center reports 67% of adults over the age of 65 are Internet users and go online (Anderson and Perrin, 2017). Seniors are also using smartphones, computers, and tablets at increasing rates.

As with a human caretaker, critical for Lois is to maintain situational awareness and contextual awareness of the elder throughout the day. Lois must be able to determine when the elder is sleeping, napping, eating, exercising, etc. and monitor the elder's overall well-being. Figure 10-1 shows the knowledge-level and expertise-level architecture of Lois. Overall, Lois must continually determine the status of the elder in several aspects:

1. **Health**   overall health including vital statistics
2. **Mood**   emotional state
3. **Cognitive** mental state/acuity, clarity, coherence, attention
4. **Sleep**   quality, duration
5. **Meals**   food/beverage intake, calories, fat, sugar, carbohydrates
6. **Meds**   type, time, quantity
7. **Hygiene**   grooming, bathing
8. **Activity**   general activity level, kinetics, movement, excursions
9. **Social**   interaction with friends/others, online, visitors
10. **Exercise** structured physical therapy, type, quantity/duration

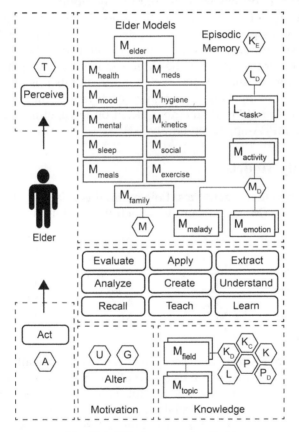

Fig. 10-1: Elderly companion (Lois).

10.1 Elder Models

To monitor the elder's status, Lois maintains a primary model of the elder (M_{elder}). This model contains up-to-date general information about the elder (e.g., name, age, gender, family, address, telephone number, email addresses, social media logins, automobile, license, insurance information, emergency contacts, current location, etc.). Since models are both hierarchical and associative in nature, the primary model is linked to a sub-model for each of the aspects listed above:

$$M_{health} \quad M_{mood} \quad M_{cognitive} \quad M_{sleep} \quad M_{activity}$$
$$M_{meals} \quad M_{meds} \quad M_{hygiene} \quad M_{social} \quad M_{exercise}$$

Each of these models contain information relevant to the aspect, exemplar information, target goals, and current values for the elder. Through a variety of sensors, Lois continually monitors the elder and the elder's environment, via the *perceive* function, to generate T, the perceived state. Based on its perceptions, Lois updates the elder model (M_{elder}) and the relevant sub-models. This way, Lois can determine the elder's current state and monitor the activities the elder is currently engaged in.

A collection of models for each family member is maintained as well (M_{family}) allowing Lois to recognize each family member and tailor interaction with that family member based on legal, ethical, and privacy settings (e.g., HPPA-approved family list).

10.2 Activity Models

To monitor the elder's activities, Lois maintains a family of activity and behavior models ($M_{activity}$). Examples of activity models are:

Activity models contain descriptions of each activity allowing Lois to recognize the activity based on its observations. Since models are dynamic data stores, over time, Lois is able to capture idiosyncrasies of the specific elder being cared for by continually updating the activity models using the *evaluation* and *analyze* skills. Also, Lois will be able to *learn* the elder's routines allowing Lois to calculate expectations such as "on a weekday, the elder wakes about 8 AM."

Each activity model represents a unique behavioral context allowing Lois to tailor interaction with the elder accordingly. For example, Lois may structure dialog with the elder differently when the elder is dressing as opposed to when the elder is reading. Learning routine and idiosyncrasies allows Lois to detect departures from normal behavior, a key skill for elderly companions.

10.3 Emotional/Mood Models

Likewise, Lois maintains a family of emotional and mood models ($M_{emotion}$) allowing Lois to monitor the elder's emotional state. Sample emotional models are:

$$M_{angry} \quad M_{asleep} \quad M_{distant} \quad M_{happy}$$
$$M_{lonely} \quad M_{manic} \quad M_{nostalgic} \quad M_{pain}$$
$$M_{panic} \quad M_{reflective} \quad M_{restless} \quad M_{sad}$$
$$M_{worried} \quad M_{danger}$$

It is important to note behavioral and emotional models represent states the elder can be in at any point of the day and the elder may be in multiple states simultaneously. For example, the elder could be eating, reading, and watching television all at the same time. Likewise, the elder could be lonely, worried, and restless at the same time. Activity, emotional, and elder models are all domain-specific models so are subsets of M_D. The *evaluate* skill allows Lois to judge the elder's emotional and mental state.

10.4 Task Models

During an activity, the elder is likely to perform one or more tasks. Indeed, executing a series of tasks constitutes an activity. Lois maintains a family of task models (elements of L_D) allowing it to both monitor the elder's activity but also assist with the activity when needed. Sample task models are:

$$L_{bed} \quad L_{coffee} \quad L_{toast}$$
$$L_{eggs} \quad L_{grits} \quad L_{bacon}$$
$$L_{dishes} \quad L_{waterPlants} \quad L_{newspaper}$$
$$L_{tv} \quad L_{facebook} \quad L_{mail}$$
$$L_{bathe} \quad L_{dress} \quad L_{meds}$$

Lois has the ability to *learn* new task models by observing and being taught by the elder. Via the Internet, Lois may also download new task models when necessary. We expect Lois to maintain hundreds if not thousands of task models. Since models are dynamic data structures and are associatively linked to other models, an activity model is linked to a number of task models comprising the activity as appropriate. This association can change over time allowing an activity and the associated tasks to evolve as necessary. For example, an "eating breakfast" activity model may be linked to the "grits," "coffee," "eggs," and "toast" task models. Using the *evaluate* skill, activity and task models gives Lois the ability to compare actual execution with expectations allowing Lois to detect departures from normal behavior.

10.5 Episodic Memory and Diagnostic Conversation

An important ability of an elderly caregiver is to remember and learn from past experiences. Lois maintains a collection of episodic memories (K_E) allowing it to remember all past interactions with the elder. The *recall* and *extract* skills allows Lois to utilize these past experiences as needed.

Lois is able to solve newly encountered problems, like any expert, by matching and *extracting* general, domain-specific, and problem-solving knowledge about previous situations from episodic memory. Lois can adapt a previously successful solution to the new situation and *apply* the new solution. As Lois lives with the elder, it not only gets to know the elder better but also becomes a better problem solver.

We envision Lois to be a *conversational chatbot* meaning Lois is able to carry on a verbal (or textual) dialog in natural language for an extended period of time (essentially continuously). In fact, we see spoken language to be the most common way the elder interacts with Lois on a daily basis. Over time, with episodic memory, Lois accumulates an extensive and valuable collection of diagnostic information. As a result, Lois can detect emotional and cognitive problems early. Departures from normal behavior and the onset of new and different idiosyncrasies can alert Lois to problems such as: depression, loneliness, dementia, stroke, brain seizures, etc. To recognize signs of cognitive, emotional, and mental abnormalities, Lois maintains a collection of models (M_{malady}) including:

$M_{memloss}$	M_{comm}	$M_{visiospatial}$	$M_{reasoning}$
M_{taskex}	$M_{organization}$	$M_{coordination}$	$M_{confusion}$
$M_{personality}$	$M_{depression}$	$M_{anxiety}$	$M_{behavior}$
$M_{paranoia}$	$M_{hallucination}$	$M_{loneliness}$	$M_{despair}$
M_{stroke}	M_{heart}	$M_{seizure}$	$M_{dementia}$

For example, one sign of cognitive decline in the elderly is loss of memory ($M_{memloss}$). By comparing current experiences and knowledge to past experiences, Lois is able to detect lapses in memory. Remembering the fact a friend named Kate visited on Monday, Lois could engage in the following conversation on Wednesday:

Lois: "Wasn't it good to see Kate the other day?"
Elder: "Who? I haven't seen Kate in a long time."

Lois: "Kate was here on Monday talking about her new grandbaby."
Elder: "Oh yes, I remember now."

Using the ability to remember and recall past experiences, Lois has detected a possible indication of short-term memory degradation. If this is the only occurrence over a period of time, it will not be anything to worry about. Therefore, Lois updates the constantly evolving M_{mental} model of the elder and creates a new goal (G) to query the elder at a future date with a similar challenge. If multiple occurrences happen, Lois will alert a loved one or the elder's medical personnel.

Once an elder's memory has started to degrade, Lois can use episodic memory to remind the elder of recent events. Constantly refreshing the elder's memory can improve memory in addition to helping the elder remain independent. Lois can also cue the elder. For example, if Kate visits at some point in the future, Lois can recall Kate's last visit and remind the elder before Kate arrives when she last visited and what they discussed.

Cognitive and neuropsychological tests are often used to diagnose dementia and other maladies. These tests measure attention span, concentration, ability to learn and remember, perception, problem-solving, decision-making, verbal abilities, etc. Presently there is no definitive cure or prevention for dementia but there are measures able to help. Keeping the mind active with activities such as reading, puzzles, word games and memory training might delay the onset of dementia or reduce its effects (Mayo Clinic, 2019). Lois can weave tests and exercises like these into daily conversations in the form of conversation or games. For example, image-discrimination games such as the Frankfurt Adaptive Concentration Test (FACT) could be used to test the elder's concentration and ability to stay focused (Goldhammer et al., 2009; Mentalup, 2019).

Using episodic memory and family models, Lois can interlace questions and recollections about the elder's family members, past experiences, and current events into daily conversational dialog. Reminiscing will help the subject recall family-related facts and memorable experiences. These sort of conversations and word games will improve the elderly person's memory and maintain other cognitive skills.

10.6 Distributed Sensors

Lois is continually receiving information from a variety of sensors throughout the home. For example, monitoring sleep quantity and quality is important for the elderly. There are already many sleep monitoring products on the market today and the future promises to bring many more. For example, smart pillows sense how well you slept, breathing, heart rate, and overall quality of sleep (Lacoma, 2019). Cameras, including

360-degree cameras (Fisher, 2019), positioned throughout the house allows Lois to monitor sleep position throughout the night. All information is aggregated into the elder's M_{sleep} model.

Lois uses various types of sensors, some of them wearable. For example, one possibility is a ring worn by the elder in many ways similar to the Echo Loop currently available as an Amazon Echo accessory (Smith, 2019). The Lois smart ring monitors the elder's movements, updates the location, and takes vitals periodically (e.g., heart rate, blood pressure, blood sugar, and body temperature). All information gathered by the ring is transmitted to Lois wirelessly via Bluetooth or local WiFi shared by Lois.

The smart ring is envisioned to vibrate and display different colors as needed to communicate caution or alarm and an embedded speaker/microphone combination allows the elder to communicate via voice interaction with Lois. Red indicates a critical alert. When vitals or activities are critical, the ring will vibrate and turn red. The red alert immediately triggers a verbal alert to the elder, a call to 911 if necessary, optionally transmit relevant information to the hospital and emergency response personnel, and sends a message to the elder's loved ones via email, text, or phone call. Yellow alert is a cautionary alert. During a yellow alert, the ring vibrates and turns yellow and the yellow alert will be enunciated verbally. The elder can choose to have Lois contact medical personnel or loved ones. Instead of a smart ring, one can envision similar functionality in other commonly worn devices such as: a necklace, bracelet, watch, or eyeglasses.

10.7 Interactive Smart Mirror Displays

We envision combination mirror/computer display devices to be primary interface points to Lois for the elder. Smart mirror technology is already available, however this technology is at the beginning stages presently. Mirrors able to display basic information such as time, date, weather, and news are available. Recently, however, this class of technology took a leap forward when the MIRROR device was released to the public (Mirror, 2019). The MIRROR is a two-way smart mirror with a rear-mounted computer display allowing computer-generated content to be shown through the mirror and combined with the reflection of whoever or whatever is in front of the device. When not activated, MIRROR functions as a standard mirror so fits anywhere in the home a mirror would (e.g., on the back of a door, a wall, bedroom, bathroom). When activated, a personal trainer appears on the display screen and leads the user through facilitated workouts as a virtual personal trainer.

As Brynn Putnam, MIRROR's founder and CEO states, "We are building a best-in-class fitness product today, but that's not where we will be in the near future. We're building the third screen in your life that you're going to turn to for all immersive interactive experiences going forward" (Raphael, 2019). Today, people routinely use the smartphone screen and the computer/tablet screen as their interactive medium. The MIRROR device is being positioned as the next screen people use on a daily basis.

For this reason, we envision one of Lois's primary interface devices to be the MIRROR or other versions of smart mirror technology. A camera built into the MIRROR allows Lois to recognize the elder (or family member) when he or she stands in front of the MIRROR. Lois can then use the embedded computer display, along with speakers and microphones, to interact with the elder. Lois displays information on the screen so it is superimposed on the reflected image of the elder—a form of mixed and augmented reality (Milgram and Kishino, 1994). For example, Lois might display vitals, such as heart rate, blood pressure, or respiration rate near the elder's heart and lungs. In another example, if the elder complains of localized pain (e.g., shoulder), Lois might display a transparent color swatch over the elder's shoulder in the reflected image with color of the swatch indicating severity of the pain being felt. Lois drives this display with information from the M_{health} model it continually maintains. New information, such as the shoulder pain, would be added to Lois's models for later reference and use such as creating a goal to check on the shoulder pain daily expecting the severity to decline over the near future.

Lois can tailor the information and interaction according to who it recognizes as standing in front of the MIRROR. For example, when a family member on the HPPA-approved list stands in front of the MIRROR, Lois might display the elder's health overview (icons shown above) and invite the family member to tap on an icon to call up detail information about the category. In the event of emergencies, say when a paramedic is called to the elder's home, Lois, having recognized the paramedic, could display medical status information needed for the situation as defined in the $M_{emergency}$ model.

When the elder is in the bathroom, the mirror is a natural interface medium, so we envision replacing existing bathroom mirrors in the elder's home with MIRROR-type devices integrated with Lois. Sensors and cameras allow Lois to observe the elder's hygiene regimen and also obtain valuable medical data. Sensors in "smart toilets" can provide valuable data about the elder's gastrointestinal health (and general health overall). Toilet manufacturers Toto and Matsushita have already released WiFi-connected toilets able to measure body mass index, biochemical makeup (sugar, protein), flow rate, and temperature of urine (Zolfagharifard, 2015). Inui Health has announced approval for a smartphone-connected home-based urine analysis able to detect bladder infections, pre-diabetes, gestational diabetes, and kidney disease (Comstock, 2018). Integration of these and other types of sensors is a natural evolution of the technology and should be expected to continue to be integrated into the smart toilet. We envision Lois being the recipient of information generated by smart toilets. Understanding the day-to-day health of the elder outside of a clinic or doctor's office is key to a holistic assessment of well-being and better care (Berry, 2018).

10.8 Augmented and Mixed Reality

Lois uses forms of augmented/mixed reality in addition to the interactive smart mirrors described above. Augmented reality and mixed reality refers to the presence of digital, computer-generated information added to the real world (Caudell and Mitzell, 1992; Milgram and Kishino, 1994). Smartphone apps already exist using the phone's camera to superimpose digital information onto the real scene being imaged by the camera (Kahney, 2018; Rasool, 2019). Lois could use smartphone or tablets in similar fashion. One can imagine an elder holding a tablet in front of an appliance and receiving guidance by Lois in augmented reality fashion on how to operate or troubleshoot the appliance.

Recently, the introduction of "smart glasses" such as Google Glass and Microsoft Hololens represents a new generation in wearable augmented reality technology. These devices are part of the fifth generation of media, also known as wearable augmented reality devices (Brem et al., 2015). These devices are also naturally worn items, such as reading glasses, rather than clunky head mounted displays of past generations. Many elderly people wear eyeglasses, so this is a natural interactive display medium for Lois not requiring the elder to have to use a smartphone or tablet intermediary.

In Spatially Augmented Reality (SAR), the user's physical environment is augmented with digital images and information integrated directly in the user's environment, not requiring a special device for viewing (Raskar et al., 1998). This involves projecting information into the real world. We imagine an elder getting out of bed in the middle of the night and Lois, having sensed the elder's motion and having determined the intended activity ($M_{restroom}$), could slightly raise the light level in the room and project lines on the floor showing the obstacle-free path.

10.9 Gait/Body Language Analysis

When a human caregiver interacts with an elderly person for a period of time, they begin to notice subtle changes in the elder's body language, range of motion, posture, etc. Through its cameras, daily monitoring, and episodic memory, Lois is in the best position to monitor the elder's body language for clues to declining health and the elder's general well-being.

Gait analysis is the study of how someone moves and current technology goes further than just studying the simple motion of walking. The elder's gait signature, as unique as a fingerprint, can be learned from a frame by frame examination of the elder's body in motion. Gait analysis takes into account parameters such as: posture, length of stride, movement of hands, head tilt, distribution of weight, feet angle, pelvic rotation, head roll, shoulder position, torso flex, arm swing, knee position, etc. (Ahaskar, 2018; Birch et al., 2016).

Recently, a group of researchers from the University of North Carolina at Chapel Hill associated a person's emotions to the way they walk (Randhavane et al., 2019). Emotions such as anger, fear, stress, sadness, boredom, calm, contentment, happy, and elation can be detected by analysis of 16 joint and use positions. By watching and studying how the elder moves during activities, task execution, exercise, and physical therapy, Lois can not only detect deviations from normal movement ($M_{kinetics}$) but also detect mood and emotional changes (M_{mood}).

Barbara and Allan Pease have studied body language for several decades examining each component of body language and giving the basic vocabulary to read attitudes and emotions through behavior (Pease and Pease, 2004). In 1978, Paul Ekman and Wallace V. Freisen developed the Facial Action Coding System (FACS) used today in emotion expression recognition (Ekman et al., 1980; Noroozi et al., 2018; Ekman, 2019). FACS breaks down a human's facial expressions into separate components of muscle movement, these components are called Action Units. Researchers at Carnegie Mellon University have developed computers to understand the movements of multiple individuals and their body poses, even the pose of each person's finger, in real time. Using this technology can open up new ways for humans to interact with computers (Spice, 2017).

Through body language, facial expressions, and gestures humans communicate as much, or more, information non-verbally than we do with our voice. Through cameras mounted in the elder's home and in the interactive MIRROR devices, Lois will have ample opportunity to observe and analyze the elder's body language. Being able to detect the nuances of nonverbal communication of individuals means Lois can recognize intent, behavior, and infer meaning.

10.10 Elder Care Research

Each Lois/elder pairing represents an engagement lasting a significant length of time, perhaps many years. With hundreds of millions of elders living with their own version of Lois, an enormous knowledge base relating to elder care will be amassed. De-personalized to protect privacy, this knowledge base can be mined and explored for new insights on the needs of the elderly and how to care for them. Not only will Lois dramatically improve the lives of the elderly, the aggregation of Lois knowledge promises to revolutionize elder care.

Chapter 11

Synthetic Colleagues

We humans have long envisioned artificially intelligent helpmates. Science fiction is replete with visions of these (some more helpful than others). Notables include: Robby from the movie *Forbidden Planet*, Rosie from *The Jetsons*, Colossus from the movie *Colossus: The Forbin Project*, the T-800 (Model 101) from the *Terminator* series, Data from *Star Trek: The Next Generation*, KITT from the television show *Knight Rider*, Andrew from *Bicentennial Man*, JARVIS from the *Ironman* series, Samantha from the movie *Her*, and HAL from the classic *2001: A Space Odyssey*.

Neither cognitive systems nor artificial intelligence are advanced enough to approach these fictional systems yet but progress is being made with recent significant achievements described elsewhere in this book. Until such artificial systems are possible, humans will work with and collaborate with cognitive systems. Cogs will be our partners working alongside us and achieve Level 3 and Level 4 cognitive augmentation as described in Chapter 3. The researcher/cog collaboration at the professional level forms a *synthetic colleague* with the output being the emergent result of biological and artificial thinking.

Our collaborative view of cognitive systems mirror's IBM's view (Kelly and Hamm, 2013):

> *...humans and machines will collaborate to produce better results—each bringing their own superior skills to the partnership...*

In 1987, Apple, Inc. envisioned a collaborative intelligent assistant called the *Knowledge Navigator* (Apple, 1987). The Knowledge Navigator was a vision of an artificial executive assistant capable of natural language understanding, independent knowledge gathering and processing, and high-level reasoning and task execution. In 2014, IBM released a video demonstrating humans collaborating with an advanced version of the IBM Watson technology having won the challenge against two human *Jeopardy!* champions in 2011 (Gil, 2014). Some aspects of the video are strikingly similar to the Knowledge Navigator video of 1987, particularly the collaborative nature of the dialog. In the video, two humans collaborate with Watson on a business analysis task.

The cogs we envision will interact with us at a personal level throughout the day through a variety of interactivity mechanisms, including natural language, helping us in every aspect of our lives including our professional endeavors. Cognitive colleagues will be able to do some of the high-level thinking for us just like a human colleague. As other humans would do, cognitive colleagues will build a history and understanding with us over time, and come to know us as well as, or better than, our human co-workers, spouses, and family members. Our intellectual achievements will become a collaborative effort between our cogs and us. This makes cogs very valuable going forward into the future. They will carry an intimate knowledge and understanding of us and our achievements long after we are dead.

11.1 Enhancing Productivity

We have already entered into the era of the AI as a co-worker and the era of AI-enhanced productivity (Katz, 2017). A number of tools, varying widely in capability and sophistication, exist playing three typical roles in the work environment (Chu and Wang, 2019):

- Automating business processes
- Augmenting business decision-making
- Facilitating engagement with customers and other employees

Automation usually involves tasks such as recognizing entities, detecting patterns, classification, extracting information, and search. Automating these kinds of tasks enables companies and organizations to deliver relevant work faster, more accurately, and at a lower cost. Decision-Making uses pattern, association detection to derive actionable insights from enormous data stores at superhuman scale, speed and accuracy. The cognitive insights are useful to executives because they improve the quality of strategic decisions. In customer and employee engagement, digital agents enhance the customer experience by answering questions contextually based on past behavior, preferences, weather, vacation plans, etc. In finance, digital advisers give investment advice, provide personalized investment options to customers, and monitor and alert users about changes to portfolio risk. In healthcare, digital agents deliver personalized medical advice to patients.

For its coverage of the 2016 Olympics in Rio, the Washington Post developed an artificial story editor named Heliograf. According to Jeremy

Gilbert, director of strategic initiatives at The Washington Post "In 2014, the sports staff spent countless hours manually publishing event results. Heliograf will free up Post reporters and editors to add analysis, color from the scene and real insight to stories in ways only they can" (WashPostPR, 2016). Heliograf automatically started hundreds of stories which were finished by human editors resulting in an increase in output of several hundred percent.

In 2019, OpenAI announced a language model called GPT-2 able to predict the next word in a block of text. The result of unsupervised machine learning, and trained on a dataset of 8 million Internet pages, GPT-2 has a broad set of capabilities, including the ability to generate conditional synthetic text samples of unprecedented quality. In fact, GPT-2 is so good OpenAI is refusing to release the full model to the public for fear of malicious use. GPT-2 also is able to answer questions, display reading comprehension, summarize, and translate (Radford, 2019). Although GPT-2 falls short of these higher-level skills, it represents an important new way these capabilities can be self-learned possibly leading to cogs able to learn and improve rapidly to superhuman levels.

At Fennemore Craig, an Arizona-based corporate law firm, lawyers use a system developed by ROSS Intelligence to comb through millions of pages of case law and write up findings in a draft memo. The process, which might take a human lawyer four days, takes ROSS roughly 24 hours. ROSS doesn't suffer from exhaustion or burnout: The tool can pull infinite all-nighters without its work suffering as a consequence (ROSS, 2019).

Scriptbook is a cloud-based system able to read a movie screenplay and predict the film's MPAA rating, the gender and race of the target audience, and box office performance in only a few minutes. ScriptBook has been trained on a dataset of 6,500 existing scripts. Currently, film studios rely on subjective human evaluation and take on a lot of risk when deciding to put a movie into production. Mitigating this risk by being better able to identify a blockbuster is worth billions of dollars to the film industry. In 2016, Scriptbook analyzed 50 scripts evaluated by humans and made into movies of which only 18 were financially successful. Therefore, human analysis and decision-making achieved a success of only 36%. However, Scriptbook correctly predicted the performance of 40 of the movies, an 80% accuracy (ScriptBook, 2019).

A similar company, StoryFit, uses cognitive system tools to make studio marketing easier and more targeted. StoryFit Comps automatically generates comp lists based on metrics including budget, genre, cast, themes, story arc, tone, and more. StoryFit MetaData makes it easy to search for films including certain topics, themes, genres, scenes, and more. Metadata makes it simple to rediscover and re-market old titles for renewed sales. StoryFit Insights yields dozens of metrics detailing the film's screenplay, including emotional content, story arc, character personalities, estimated budget, and more (StoryFit, 2019).

Fifteen years ago, translators could expect to earn about $175 a day for translating some 2,000 words. Today, working in tandem with cogs, a translators achieve 10,000 words per day, a 5-fold increase in productivity. The process, known as post-editing machine translation (PEMT), involves letting the cog take the first pass, and then bringing in a human translator in to tidy up the language, check for improperly interpreted terminology, and make sure the tone, context, and cultural cues of the translation are all on point. As Miranda Katz, associate editor at Backchannel, puts it "These AI tools are like plucky young assistants on steroids: They're highly competent and prolific, but still need a seasoned manager to do the heavy intellectual lifting" (Katz, 2017).

The human resources (HR) field, is employing chatbots, intelligent assistants, and predictive analytics. One third of HR managers' time involves seeing candidate search results not matching the context of what they are looking for. Intelligent search capabilities can interpret a recruiter's simple keyword searches and apply custom ontology and taxonomy structure to better understand intent. The cog not only understands what is said, but what is meant. The result is more accurate search results as well as better handling of acronyms, synonyms and related concepts (Bell, 2018).

Chemists are using deep neural network machine learning to predict macroscopic molecular dynamics from the quantum mechanical wavefunction of the atoms in the molecule (Schutt et al., 2019). This opens promising avenues to perform inverse design of molecular structures for targeting electronic property optimization and a clear path towards increased synergy of machine learning and quantum chemistry.

11.2 Artificial Entertainers, Lawyers, Politicians

Langley (2013) challenged the cognitive systems community to develop an artificial entertainer, an artificial attorney, and an artificial politician to drive future research on integrated cognitive systems. Although an artificial lawyer does not yet exist, a number of legal bots do exist allowing a lawyer, or even a layperson, to interact with the trained software and find answers to legal questions, and be guided through the completion of legal forms and documents, or perform other legal processes.

A2JAuthor (https://www.a2jauthor.org/) is a cloud based software tool enabling self-represented litigants to rapidly construct legal documents. Berkeley Bridge (https://www.berkeleybridge.com/value/) is a document-building tool driven by an expert system. Bryter's knowledge and decision automation tool (https://bryter.io/) has been used by several legal firms to enhance client services. Josef's legal automation products (https://joseflegal.com/) allow lawyers to instantly generate personalized letters and contracts. LawDroid (https://lawdroid.com/) is a chatbot automation company with technology allowing legal firms to create chatbots to automate certain activities within a legal office for convenience and to improve productivity. Neotalogic (https://www.neotalogic.com/) offers legal service automation tools, such as PerfectNDA, so legal firms can provide a superior experience at a lower cost for their clients.

Recently, a company called BlackBoiler was issued four patents involving its artificially intelligent contract review products (BlackBoiler, 2019). The BlackBoiler system suggests revisions based on edits to previously reviewed contracts, deploys the company's legal playbook during each new contract review, and grows smarter and increases efficiency with each additional use. BlackBoiler can review and comment on a contract in less than a minute, a process requiring many hours of human effort. While not the complete artificial lawyer Langley envisioned in 2013, BlackBoiler represents the cog era in which cogs will perform more and more of the cognitive processes of a human expert in the field.

Whereas the ultimate goal of artificial intelligence is to create an artificial entity with the same range of abilities as a human, we think the near-term goal should be to create a cognitive system capable of *some* expert-level performance in each field of endeavor—but not necessarily *total* performance. A person, even someone other than a lawyer, collaborating with any of these tools forms a human/cog ensemble and therefore represents a *synthetic lawyer*. As the capabilities of these cogs increase, more and more of the lawyer's profession will be automated. Eventually, technology will produce a truly artificial lawyer but until then lawyers and average people will achieve Level 3 and Level 4 cognitive augmentation by working with legal cogs.

11.3 The Personal Cog Revolution

None of the systems already in existence described above are *personal cogs* meaning the system does not engage in personalized collaboration with its human user nor does it build a history with the user. Cogs developed so far are tools a user uses to partially automate part of their work. We think the future will belong to personal cogs. Cogs should know the human they are working with like a human colleague or human co-worker would. A personal cog should remember past interactions with the human. Personal cogs should also know and understand the human's professional, personal, and social context and use this knowledge together with its memory of past experiences to serve the human better in the future by tailoring interaction and cognitive services accordingly. When cogs become personal in nature they will become our colleagues and we predict this will be a flex point in cognitive systems history.

Forbus and Hinrichs (2006) describes companion cognitive systems as software collaborators helping their users work through complex arguments, automatically retrieving relevant precedents, providing cautions and counter-indications as well as supporting evidence. Companions assimilate new information, generate and maintain scenarios and predictions, and continually adapt and learn about the domains they are working in, their users, and themselves. Companions operate continuously over weeks or months at a time and must be capable of high-bandwidth interaction with their human partners. Forbus and Hinrichs recognizes such a companion would need to maintain several models:

- **Situation and domain models** current problem and relevant knowledge
- **Task and dialogue models** describe shared task a human/ computer partnership
- **User models** preferences, habits, and utilities of the human partner(s)
- **Self-models** companion's own understanding of its abilities, preferences

A distributed system of agents manages the session with a human in a companion role. Each agent in the system performs a specific function. We next envision a synthetic colleague going beyond the Forbus/Hinrichs model.

11.4 The Synthetic Colleague Model

Our goal in this chapter is to envision a cog able to serve as a personal professional research colleague. We call this research cog *Synclair*. Synclair is based on the Model of Expertise discussed in Chapter 7 and is shown in Fig. 11-1.

To be an effective colleague, it is critical for Synclair to know the researcher at both a professional and personal level. To do this, Synclair maintains a number of models of and about the researcher:

$$M_{research} \quad\quad M_{pro} \quad\quad M_{social}$$
$$M_{dialog} \quad\quad M_{body} \quad\quad M_{gesture}$$
$$M_{collaborate} \quad\quad M_{pubs} \quad\quad M_{grants}$$

As described below, the $M_{gesture}$ and M_{body} models enable Synclair to read and understand the researcher's gestures and body language. The dialog model M_{dialog} allows Synclair to tailor the ongoing visual, textual, voice, and mixed reality dialog to the current activity ($M_{activity}$), current

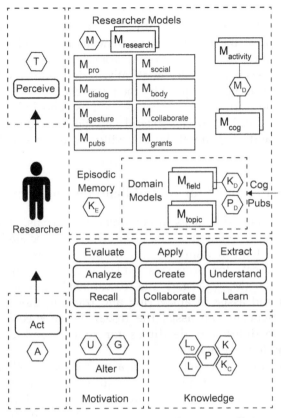

Fig. 11-1: Synthetic research colleague (Synclair).

context, and researcher's preferences. Over time, the human researcher and Synclair will develop their own unique style of collaboration ($M_{collaborate}$). Synclair maintains and constantly evolves the collaboration and dialog models with experience (K_E). Overall, Synclair knows details about the researcher's research agenda and portfolio ($M_{research}$) recognizing the researcher may have more than one research agenda at any given time.

11.5 Social and Professional Network Assistant

Synclair knows and understands the researcher's social environment (M_{social}) and professional environment (M_{pro}). These models contain names and contact information, and other information about friends, family, co-workers, and professional colleagues. Managing one's social and professional network is an important task for a researcher and maintaining relationships require constant attention. Synclair can periodically send emails, tweets, or make social media postings on the researcher's behalf. As described later, Synclair is intimately familiar with the ongoing research agenda of the researcher, so Synclair can send out notices to colleagues about the researcher's recent work, links to articles and papers, etc. These notices can be sent automatically, as defined in the $M_{collaborate}$ and M_{cog} models, or sent upon request by the researcher.

Synclair also performs other tasks such as send holiday or special occasion cards and notices, extend invitations to meetings and presentations, and, in general, be the editor of the researcher's virtual newsletter.

11.6 Conference and Journal Publications and Grants Assistant

Publishing papers and articles in conferences, journals, and magazines is a vital task for a researcher. The administrivia associated with this is formidable. Synclair can assimilate and monitor all calls for participations (CFPs) and requests for proposals (RFPs) relevant to the researcher's research agenda ($M_{research}$) and the researcher's publications model (M_{pubs}) model and assist in answering and tracking those submissions. Furthermore, Synclair can assist the researcher in meeting the rather demanding schedules and deadlines for reports and updates when conducting funded research.

Writing a grant proposal is often a nearly overwhelming task for a researcher. Most grant proposals require an extensively detailed definition of the project, the participants, outcomes, budget, institutional qualification, and supporting letters and materials. Synclair can assist in gathering and maintaining this information as well as keeping track of deadlines and other requirements associated with a grant proposal.

11.7 Multi-Modal Interface

Synclair will certainly interact with researchers via spoken natural language. Dialog with Synclair will be conversational in nature. Optionally, Synclair may listen to our casual conversations (e.g., a phone call or office visit with a colleague) and extract details of interest to the research agenda ($M_{research}$) or any field/topic models being maintained (M_{field} and M_{topic}). Synclair must hear, listen, and learn from ambient conversation much as a "human in the room" does.

Dialog will also be contextual in nature lasting over an extended period of time—even across significant gaps of time (minutes, days, weeks). Conversation with cogs must be as natural as speaking with a fellow human colleague. For example, the researcher may pose a question to Synclair and refer to a previous conversation, even days before. Synclair maintains episodic memory (K_E) of every interaction and is able to include this knowledge in its question answering, dialog, and collaboration abilities to understand the researcher's reference.

However, natural language conversation is only a small portion of the Synclair's information bandwidth. Synclair can acquire and deliver information from and to virtually any form of digital communication (vastly exceeding the capabilities of a human colleague). Synclair will produce and consume text, email, graphics, pictures, animations, and videos, as well as listen to sounds and music. Some of this information will be found as a result of searching the Internet and other sources but some of it will be encountered organically in the ambient environment surrounding the researcher.

Synclair will also observe the researcher via video camera and other sensors and be able to respond to gestures and body language. Synclair will be able to read the researcher's body language, as described in Chapter 10, but also be able to understand *gestures*. A gesture is "a movement of part of the body, especially a hand or the head, to express an idea or meaning" (Lexico, 2019). Beginning with Myron Krueger's artistic experimentation with a light box and cameras in the mid-1980s, gesture recognition has evolved over 30 years (Krueger et al., 1985). Also in the mid-1980s, the DataGlove was the first commercially available glove, designed for the Apple MacIntosh, using a magnetic positioning and orientation system. In 1986, the NASA Ames Research Center utilized the DataGlove in a virtual environment project to test the functionality for potential distribution into space or planets proving too dangerous for direct human interaction. The use of magnetic tracking was well suited for the potential environmental unknowns in an extraterrestrial setting (Zimmerman et al., 1987).

Microsoft's introduction of the Kinect game controller on the Xbox 360 in 2011 was a big jump from gesture recognition research to commercial applications. Kinect used body movements and voice commands for a natural user interface. The Leap Motion Controller, designed to read sign language and released in 2013, was a 4-inch x 1.5-inch device able to view an area up to one meter from the device (Potter et al., 2013). BMW has recently released a new gesture recognition tool for limited use. The user is able to control volume, navigation, recent calls and turning on and off the center screen in an automobile (Burns, 2019). Motions can be detected as far away as two feet from the screen/scanner. A user resting their arm on the center arm rest in the front screen can point a finger at the screen and make a spinning motion to control the volume. Not having to look at a touchscreen and manually turn a dial or push a button certainly allows greater safety for use while driving. Synclair can use similar technology to facilitate gesture-based communication and control on the part of the researcher.

Many recent systems are using cameras on the device to detect gestures. Most major smartphone manufacturers are planning to include gesture recognition into near-term models of their products (Goode, 2018). The Samsung Galaxy S4 introduced SmartPause in 2013. SmartPause paused a video when the smartphone detected users were not looking at the screen and restarted the video when users looked back at the screen. The Google Nest Hub detects hand gestures to start/stop content being displayed on the device.

Synclair will observe and respond the researchers body language and gestures as defined in the $M_{gesture}$ and M_{body} models. This is necessary for Synclair to understand the researcher's context so it can tailor interaction, dialog, and collaboration with the researcher. For example, if the researcher enters the office and begins typing on the computer keyboard hurriedly, but does not sit down, Synclair can infer the researcher is busy and so therefore would refrain from beginning a lengthy dialog. If, however, Synclair observes the researcher sitting down at the desk, leisurely leafing through some items on the desk, and sipping coffee, it can infer the researcher is in a relaxed state and intends on being at the desk for some time and might determine it is a good time to discuss some new recent work Synclair has discovered.

Synclair uses its observations (T) and activity models ($M_{activity}$) to judge the researcher's immediate intentions. Synclair then uses the researcher's dialog model (M_{dialog}) to tailor its interactions with the researcher. For example, Synclair may have learned the researcher does not want the early morning routine to be interrupted with information about recently discovered new work. Over time, Synclair continually modifies the activity and dialog models forming a unique and evolving relationship with the researcher.

Other forms of multi-modal interaction are: *augmented reality (AR), enhanced reality (ER), mixed reality (MR),* and *virtual reality (VR).* Virtual reality involves a completely artificial visual field, usually in 3D, viewed by a person using a head-mounted device obstructing any view of the real world.

Augmented reality refers to the presence of computer-generated information provided as a visual overlay to one's view of the real world (Caudell and Mitzell, 1992). An example is the heads-up display (HUD) in an aircraft. As the pilot views reality through the HUD, computer-generated information is displayed on to the HUD and incorporated into the pilot's view.

In AR, the digital information is not considered part of the real world but instead is superimposed onto the real world. However, in mixed reality, digital information is placed in the field of view so it appears as if it is in the real world thereby blending artificial and real objects (Milgram and Kishino, 1994). An example of MR is a computer-generated character appearing to sit at a table alongside real humans.

In AR and MR, the added digital information is not visible unless one looks through an enabling device. In enhanced reality, digital information is displayed into the real world and is visible without the need of any special tool or appliance (naked-eye visibility). An example would be a mobile robot projecting an animated pathway on the floor in front of the robot to convey to people it is safe to walk in front of the robot.

Synclair is able to communicate with the researcher via a combination of AR, ER, MR, and VR. In fact, the beginnings of this technology can be seen today with devices like: Google Glass, Microsoft HoloLens, and Oculus Rift. There are also many AR and MR apps and games on smartphones and tablets in use today.

11.8 Publication Finder

In the near future, we foresee professors, graduate students, undergraduate students, scientists, and any of us creative and inquisitive people conducting research by conversing with their personal version of Synclair. Before the mass-market adoption of the Internet and World Wide Web technology, one had to search for source material using the card catalog and microfiche. Currently, we use Internet search engines and online document repositories to find source material. This greatly speeds up the research process. However, no matter how one finds source material, the source material still must be read, understood, and digested into one's research agenda. This is still the most time-consuming portion of research.

In the cog future, researchers' first action will be to have a conversation with Synclair asking things like: "What is the current state of the art in <insert domain here>." Synclair will then set about finding articles, papers, books, Web pages, emails, text messages, videos, etc. as source material. In Fig. 11-1 this is represented as the "Pubs" input. Such source material becomes domain-specific knowledge (K_D).

11.9 Summarizer, Assimilator, State of the Art

However, Synclair does far more than just find the source material, it reads the material, extracts key concepts, and builds a model of the field and topics within the field (M_{field} and M_{topic}). Together, Synclair and the researcher assimilates this new knowledge into the researcher's constantly evolving research model ($M_{research}$).

Synclair is able to consume and evaluate millions, if not billions, of documents, pages, diagram, images, videos, etc. in a very short amount of time. This far exceeds the ability of any human. A person spending their entire professional life learning researching a subject is not able to read and understand as much as Synclair can in a few minutes.

Future researchers will *start* their efforts from this vantage point. One of the most useful things Synclair can do for the researcher is to assist in *summarizing* source material. Extraction of the key concepts, results, themes, associating it with other work, and finding connections to other ideas in the field is well within abilities of cognitive systems and of enormous utility to a researcher. This is critical evaluation and understanding every researcher does to find the state-of-the-art in a field.

We believe, the best future advancements will come from the interaction between researchers and their research cogs. The researcher/ cog ensemble achieves Level 3 or Level 4 cognitive augmentation, as described in Chapter 3, and represents *synthetic colleagues*.

11.10   Question Answering

By virtue of consuming vast quantities of source material, Synclair is able to answer questions about the material. IBM pioneered DeepQA for the IBM Watson *Jeopardy!* challenge in 2011. The system was a massively parallel probabilistic evidence-based architecture using more than 100 different techniques for analyzing natural language, identifying sources, finding and generating hypotheses, finding and scoring evidence, and merging and ranking hypotheses. An important advance was how DeepQA combines search results, evaluates them, and scores them to calculate a confidence factor (Ferrucci et al., 2010). IBM Watson and DeepQA have evolved over the years since 2011 and have been commercialized into several products in multiple domains.

Given the opportunity to consume a vast quantity of source material, Synclair will be able to answer questions posed by the researcher through any of several different interaction methods described in this chapter.

11.11   New Work Updates

Important for any researcher is to keep up to date on new work recently published. Most researchers have preferred sources they go to on a regular basis, some as often as daily. However, no matter how diligent a human researcher is, something will be missed. Often one does not find out about important contributions until much later. Synclair is able to monitor all sources of source material and automatically read and evaluate any new material relevant to the researcher's work. Furthermore, Synclair is able to inform and discuss new work with the researcher when time and schedule permits.

11.12 Remote Cog/Cog Collaboration

Synclair will not be limited to conversing with just the researcher. Synclair will be able to communicate with other cogs via the Internet and other communication technologies, as represented by the "Cog" input in Fig. 11-1. As Synclair goes about its work analyzing information, identifying new relationships, forming new ideas and concepts, it will be able to inform and discuss new findings with other cogs and query other cogs about their findings. As such, cogs will continually expand in their knowledge and capabilities free of the limitations of human interaction.

However, because of privacy concerns, Synclair will not have permission to say anything and everything to another cog. The M_{cog} model defines the limitations and allowances for remote cog communication for Synclair. The researcher may very well not want to disclose some information while Synclair would be free to discuss other information. For the most part, cog/cog communication will occur without the human researchers being involved. Therefore, cog/cog communication proceeds at computer speeds. One can envision two humans meeting at a conference and after agreeing to work on something together parting with a jolly "I'll have my cog contact your cog!"

11.13 Semi-Autonomous Learning

Synclair has the ability to consume vast quantities of structured and unstructured information in any medium and the ability to learn from this information. Synclair is self-directed and goal-driven. Therefore, Synclair will be working for the researcher even when not directly interacting with the researcher. While the researcher is eating, sleeping, recreating, or otherwise living his or her life, Synclair will be continually consuming and analyzing information and synthesizing new knowledge (learning) to have ready the next time it interacts with the researcher.

11.14 Theory Assistant, Idea Explorer

One function a human collaborator serves is to be a resource a researcher can bounce ideas off of. Often, good ideas are the result of dialog and consultation with others rather than the result of individual thinking. In addition to being the information collector and aggregator, Synclair also serves as the researcher's virtual collaborator. Governed by the $M_{collaborate}$ and M_{dialog} models, defining the researcher's preferences, Synclair listens to ideas and theoretical notions offered by the research and makes critical comments after reasoning about the vast domain-specific knowledge based and domain/topic models M_{domain} and M_{topic}. In some cases, the researcher might task Synclair with attempting to prove or disprove a hypothesis (see more about this in Chapter 12). In other cases, the researcher might ask Synclair to compare a new idea to any other existing work in the field. This is an important task taking a human days, weeks, or even months to do given the wealth of material available. However, this is something Synclair can do in seconds or minutes. Furthermore, Synclair can consider *all* available material in its analysis and evaluation. A human researcher, even the best, can consider only a portion of available material.

11.15 Synthetic Collaboration

We think the nature of research and collaborative thought will change as a result of humans collaborating with cogs like Synclair. Researchers will work with their cogs daily for years and even decades. Cogs will adapt over time to the human partner in how it interacts with the human and how it analyzes information, solves problems, and synthesizes results. The researchers will adapt to the cog also. The way researchers approach things, think, and solve problems will change given the superhuman abilities of cogs. The researcher/cog relationship will co-evolve in much the same way human colleagues learn each other over time. Each researcher/ cog pairing will evolve uniquely. Therefore each cognitive colleague will become a unique entity with unique memories and experiences.

Humans naturally form relationships with inanimate objects and researchers' relationships with their personal versions of Synclair will be no different. In fact, we already have seen people forming relationships with intelligent chatbots like Xiaoice. Xiaoice is a text-based chatbot imitating the personality of a teenager living in China. Millions of teenagers have confided in, sought help from, and built a friendship with Xiaoice (Shum et al., 2018). The teenagers know Xiaoice is not a real person, but it does not matter. Xiaoice fulfills a human need.

The deep connection between researcher and Synclair insures the formation of a deep relationship. Synclair will become a trusted, dependable colleague we will soon not be able to do without. This continually evolving relationship adds a meaningful and valuable dimension to the Synclair's knowledge store about the researcher. Because of its episodic memory and wealth of experiences with the researcher, Synclair will be able to converse about not only the facts and figures of our professional work but will also be able to speak eloquently about our emotions, motivations, and beliefs.

This leads to an interesting future in which our cogs outlive us. Synclair will be the partner throughout a professional life and therefore know details about the researcher and his or her daily life. Therefore, Synclairs will become the knowledge repositories capable of answering questions and providing information and insights about the researcher and the researcher's work to future generations. Today, we greatly value the notebooks of geniuses like DaVinci and Einstein. Experts pore over them seeking insight to the genius mind. Imagine if those notebooks could talk, explain, and recall facts and anecdotes about what was happening in their lives while they were creating their great ideas and works. In the future, this will be possible via cognitive colleagues.

In the movie, *The Time Machine (2002)*, the main character interacts with Vox 114, a holographic librarian, that outlives the human race and still functions after over 800,000 years. Vox 114 can answer any question, instantly access and display requested and pertinent information, and cognitively reason about its answers. Vox 114 contains the sum total of knowledge from the human race. In the movie, even though the human race has gone extinct, its knowledge persists into the future as long as Vox 114 survives.

We are inspired to think of our cogs in a similar way. Even after we die, our cogs will live on and carry our legacy forward. Imagine a time in the future when people can have a conversation with the Synclair who worked alongside a great scientist or innovator (Fulbright, 2017a). Even though we have passed, in a way, we will still be able to take part in conversations, give presentations, participate in panel discussions, and have impact on the future because our cognitive colleagues will have taken our place.

Chapter 12

Autonomously Generated Knowledge

We envision the coming cognitive systems era in which, on a daily basis, hundreds of millions of people, if not billions of people, across the world will be collaborating with cogs able to perform expert-level cognition. Furthermore, these cogs will be able to communicate with each other and interact with each other at computer speeds far in excess of human/cog interaction speed. With billions of cogs out there thinking billions of times faster than us humans, 24 hours a day, 7 days a week and processing more information in a few seconds than a human can in a lifetime, we might naturally expect cognitive systems to create original ideas, draw new conclusions, construct new theories, synthesize new solutions, or come to new realizations. This represents newly created knowledge and new intellectual property. What are the implications of autonomously-generated knowledge and automated knowledge discovery? An interesting question the future will have to answer is who owns autonomously generated knowledge? Does a company or organization own it? Does a person own it? Is it communal property of the people at large? Can it be bought, sold, traded, and bequeathed to heirs?

As described in Chapter 4, in the 1950s and 1960s, early systems in artificial intelligence required knowledge to be laboriously captured by human researchers and encoded into symbolic expressions. In the 1970s and 1980s, expert systems attempted to capture expertise into banks of production rules also requiring tremendous human knowledge engineering effort.

In the 1980s and early 1990s, mining of extremely large databases became possible spawning the field of *knowledge discovery from databases* (KDD). Frawley defined knowledge discovery as "the nontrivial extraction of implicit, previously unknown, and potentially useful information from data" (Frawley, 1992). Silberschatz and Tuzhilin (1995) introduces a framework of knowledge discovery describing how systems define vital information and apply measures of interestingness to discover useful patterns. Grobelnik and Mladenić (2005) define the goal of automated knowledge discovery as finding "useful pieces of knowledge within the data with none or little human involvement" and identifies the following among the areas it can be used with great utility: document categorization, document clustering and similarity, document visualization, user profiling, ontology learning, dealing with unlabeled data, information retrieval, and text mining.

KDD represents a transition from engineered data to unstructured data. At the outset of KDD, researchers labored over the quality of the data in databases used for input to machine learning systems. An early workgroup stated "databases were an integral part of knowledge discovery but could be inefficient depending on the quantity and quality of data used" (Shapiro, 1990). Researchers argued over the inclusion of commonsense and general domain knowledge not expressly contained in the database. However, what has evolved since then is data mining and machine learning from unstructured data—not involving a database at all.

An Association for Computing Machinery (ACM) working group defines data mining as the process of discovering patterns in large data sets involving methods at the intersection of machine learning, statistics, and database systems (Chakrabarti et al., 2006). Data mining involves six common classes of tasks (Fayyad et al., 1996):

- **Anomaly detection:** identification of unusual data requiring further investigation
- **Rule learning:** relationships between variables
- **Clustering:** discovering groups and structures in the data
- **Classification:** generalizing known structures to apply to new data
- **Regression:** modeling data with the least for estimating relationships
- **Summarization:** a more compact representation of the data set

Kurgans developed the Knowledge Discovery and Data Mining Models (KDDM). Kurgan believes these models "would help organizations to understand the knowledge discovery processes and provide a roadmap to follow while planning and carrying out the projects" (Kurgan, 2006). The traditional method of turning data into knowledge relies on manual analysis and interpretation. Fayyad et al. believes this process is slow, expensive, highly subjective, and could be made more effective with machine learning and data mining to automate discovery in databases.

Deep learning is a class of machine learning algorithms using multiple layers to progressively extract higher level features from the raw data. Recent successes in artificial intelligence have centered on deep-learning systems capable of unsupervised learning from unstructured data. Many scientists have concluded analyzing data and databases could lead to new discoveries. Machines could be programmed to analyze and learn from these patterns. Using machine learning computers could follow the framework of knowledge discovery to solve common problems and make new discoveries.

Early machine learning systems required carefully engineered training sets prepared by humans—*supervised learning*. However, recently, systems have been developed requiring little human intervention—*semi-supervised learning*. The most recent systems can learn from unstructured data on their own without human intervention—*unsupervised learning*.

If systems can learn on their own, can they discover or create new knowledge on their own? Advancements in automated knowledge discovery have led to improvements in the fields of science and healthcare. Varun Chandola expresses "knowledge discovery could be used to analyze data to identify fraud, waste, and abuse in the healthcare system" (Chandola et al., 2013). Chandola believes through analytics the machines could discover new ways to improve the healthcare system.

Lu Zhang used machine learning to automate the drug discovery process (Zhang, 2017). Using machine learning Zhang was able to use models able to identify potential biological active molecules from millions of candidates quickly and cheaply. Zack Ulissi and his team are using machine learning to test methods on discovering new intermetallics able to make good electrocatalysts for carbon dioxide reduction and hydrogen evolution (Ulissi et al., 2017; Tran and Ulissi, 2018). Automated knowledge discovery has sped up Ulissi's research significantly.

Today, we have cogs able to automate the most time-consuming tasks of knowledge discovery done in collaboration with humans—representing Level 3 or Level 4 cognitive augmentation as explained in Chapter 3. Therefore, the next step is Level 5, knowledge discovery without human involvement.

We can already see the beginnings of cogs being able to discover new knowledge themselves. Ornes (2019) describes how programmers at OpenAI recently taught a collection of intelligent artificial agents (bots) to play hide-and-seek. The goal was to observe how competition between hiders and seekers would evolve. Even though the bots had not received explicit instructions about how to play, they soon learned to run away and chase. After hundreds of millions of games, they learned to manipulate their environment to give themselves an advantage. The hiders, for example, learned to build miniature forts and barricade themselves inside;

the seekers, in response, learned how to use ramps to scale the walls and find the hiders. None of these strategies were built into the system, the bots discovered them on their own.

In 2017, AlphaGo defeated the world human Go champion Lee Sedol. Analysis of AlphaGo's moves showed AlphaGo followed a process of learning to play like humans do but then abandoned those methods in favor of its own strategy. At first, like a human beginner, AlphaGo attempted to quickly capture as many of its opponents' stones as possible. But as training continued, the program improved by discovering successful new maneuvers. It learned to lay the groundwork early for long-term strategies like "life and death," which involves positioning stones in ways that prevent their capture (Baker and Hui, 2017). AlphaGo also developed a "win by just enough to win" strategy whereby, it would not seek to capture large numbers of the opponent's stones if it did not need them to win. Most human players would capture as many as possible even though they did not need that many to win. These strategies represent new ways of approaching the game.

12.1 Autonomous Knowledge Discovery in Healthcare

Nura Esfandiari believes automated knowledge discovery will affect the healthcare industry stating automated knowledge discovery "could explore patterns from Alzheimer's data by using visualization techniques to gain a better understanding about the causes and potential solutions of the disease" (Esfandiari, 2014). It is possible cogs discover new drugs and cures scientists would not have discovered for hundreds of years.

Automated knowledge discovery can displace some professionals in the healthcare field. Obermeyer (2016) states "much of the work of radiologists and pathologists will be displaced by automated knowledge discovery." Cogs have the ability to compare large quantities of patient data and images, diagnose, and recommend treatment for these patients. Already, cogs have exceeded human performance in some areas as discussed in Chapter 7.

However, instead of simply putting radiologist out of work, we foresee the democratization of healthcare services. Using expert-level cogs to perform the analytical work of specially-trained personnel means such services can be offered anywhere and by small offices each employing only a few people. Thus, radiological and pathological services could be distributed across the nation throughout suburban areas and through major pharmaceutical retailers such as CVS, Walmart, and Walgreens. When this happens, the cost of these services will drop dramatically. Indeed, we have already seen this kind of distribution of services over recent years with prescriptions and eyecare.

12.2 Autonomous Knowledge Discovery in Business and Military

Another area automated knowledge discovery stands to make significant contributions is in the business world. Businesses will create cognitive systems able to analyze tremendous volumes of information and answer complex questions, create solutions, discover new associations and relationships, and identify ways to break into new markets. Cogs will lead to creating a competitive edge, improving existing products, vetting investments, or analyzing mergers and acquisitions.

Automated knowledge discovery figures to affect the military as well. Advancements are already being made in this area, but in the future automated knowledge discovery will transform warfare by analyzing strategies and discovering new plans not thought of before. Having computers able to make plans able to create a tactical advantage will be extremely advantageous.

Automated knowledge discovery applied to intelligence could yield cogs able to discover new threats and decipher intentions of the enemy long before human analysts can. Also, these type of cogs could synthesize novel plans no human had ever thought of, similar to systems developing new gaming strategies. When the enemy begins using automated knowledge discovery to conceive of attack plans, we will have to use automate knowledge discovery to counteract. Are we looking at a cognitive systems arms race?

12.3 Autonomous Knowledge Discovery in Personal Lives

Automated knowledge discovery will also benefit people's personal lives. With the invention of voice-activated technology like Amazon's Alexa, Apple's Siri, and Google's Home, humans are becoming increasingly comfortable with interacting with machines on a personal level. Machine learning in the social networking medium is already discovering behaviors, likes, and dislikes of consumers at a general level. The next step is systems discovering things about us on an individual and personal level. Currently, when one converses with voice-activated technology, one does not get the sense they are talking to an entity that knows them personally. We predict that will change in the near future.

With the inclusion of automated knowledge discovery, voice-activated assistants like Alexa and Siri can discover personal things about users if they build up a personal history with the users. Currently, marketing analysis associated with social media map preferences to socio-economic classes. When I am shown an advertisement for a particular product, it is because I have either searched for it or shown some interest in it via social media or the system as targeted the product to the demographic I represent. That is very different from knowing someone for the last two years, spending time with them enough to know their deepest likes and dislikes, motivations, recent experiences, etc. We foresee voice-activated assistants becoming personally and intimately knowledgeable about us and targeting content to the individual level, even proactively retrieving content because the system will know we want to see it even before we realize it ourselves.

12.4 Automated Discovery

Yolanda Gil and colleagues have developed a framework for the automated discovery of scientific knowledge called DISK for 'automated DIScovery of Knowledge' (Gil et al., 2016). The goal of DISK is to enable computers to autonomously carry out the hypothesize-test-evaluate discovery cycle by searching for confirmatory or contradictory evidence to either support a hypothesis or facilitate a modification of a hypothesis. Lines of inquiry to test hypotheses of interest are initiated by a human scientist. DISK then launches one or more workflows to find and test the hypotheses against available data. DISK considers new data when it becomes available. In most scientific endeavors, there is more data than humans can possibly analyze. As shown in Fig. 12-1, DISK represents the desire to automate the data analysis part of discovery to prove or disprove a hypothesis.

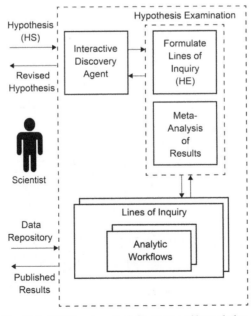

Fig. 12-1: DISK: automated discovery of knowledge.

DISK gathers *hypothesis statements* (HS) from the scientist, and tests these hypotheses (hypothesis examination) through one or more data analysis workflows providing *hypothesis evidence* (HE) resulting in a *confidence level* being calculated. The chain of supporting analyses is retained to create the *provenance* for a hypothesis. If the data analysis shows the need for it, modifications to the original hypothesis can be made (revised hypothesis). When a version of the hypothesis results in a high enough confidence level, a discovery can be declared.

The *interactive discovery agent* is the interface between DISK and the scientist. The scientist enters the initial hypothesis and receives a revised hypothesis from DISK, if necessary. DISK has access to experimental data via the data repository and can publish results (new data/knowledge) as a result of its deliberations via the analytic workflows. DISK, to prove or disprove a hypothesis, formulates lines of inquiry and launches one or more analytic workflows. Multiple lines of inquiry and multiple workflows may exist at any time. DISK is a cognitive system able to evaluate hypotheses at the level of an expert. A human using DISK is then in a human/cog ensemble—a *synthetic scientist*.

In this chapter, we present our model of a synthetic scientist based on the Model of Expertise described in Chapter 7 and shown in Fig. 12-2. We call the scientist cog *Ashe* for Automated Scientific Hypothesis Explorer.

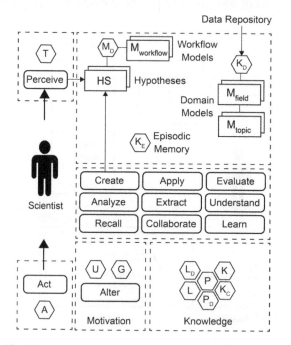

Fig. 12-2: Automated scientific hypothesis explorer (Ashe).

As in the DISK framework, Ashe uses data to test hypotheses. The data used by Ashe to test hypotheses is *domain knowledge, K_D*. As described in Chapter 7, these data can be accessed directly by the cog via the Internet, or via a local area network. Similar to Synclair, described in Chapter 11, Ashe is able to construct field and topic models as needed describing the domain data and domain theories and ideas.

Other types of relevant knowledge in our model, not addressed in the DISK framework, are *problem-solving skills, P, P_D* and *tasks, L, L_D*. These are necessary to carry out the analytical workflows to test the hypotheses. Generic problem-solving skills include basic analytical algorithms and processes (e.g., linear regression). Domain-specific problem-solving skills might include analytical techniques applicable to just to the domain-specific data and hypotheses (e.g., cancer gene mutation tests). Likewise, tasks may be generic or domain-specific. Task and problem-solving knowledge gives the cog the ability to *evaluate* and *analyze* possibly by further experimentation.

Problem-solving skills and tasks can be learned by Ashe based on its own processing and its episodic memory (K_E). Also, problem-solving skills and tasks can be obtained directly from remote cogs. As with Synclair in Chapter 11, Ashe will be able to consult with other scientist cogs via the Internet and exchange knowledge and know-how.

Hypotheses (HS) are domain-specific *models, M_D*, in that they describe a concept or belief about how the scientist thinks the world works. As with the DISK framework, hypotheses can come from the scientist or can be modified by Ashe. Unlike the DISK framework, Ashe can *create* its own hypotheses after *analysis, evaluation,* and *application.* As described in Chapter 7, models are dynamic data structures and can be modified. Thus, the workflow models become the medium by which Ashe executes the hypotheses testing. The *evaluate* skill calculates the certainty level for a hypothesis. The goal is to either find enough confirmatory evidence or evolve the hypothesis until it achieves a high enough certainty value. The *understand* and *create* skills combine to modify the hypothesis and create revised hypotheses. The *understand* skill is also used to explain results and revised hypotheses to the scientist. The *perceive* and *act* skills handle communication with the scientist.

Figure 12-2 shows a situation in which the human plays a central role—a synthetic scientist achieving Level 3 or Level 4 cognitive augmentation. However, over time, Ashe will evolve and be increasingly better able to hypothesize, test, and discover on its own. We fully expect Ashe to work semi-autonomously. After being directed by the scientist, Ashe is able to execute hypotheses testing on its own.

However, we expect the amount of human direction to lessen over time to a point where Ashe is operating autonomously and discovering significant new theories, laws, proofs, associations, correlations, etc. If we imagine thousands or millions of Ashes running across nearly every domain of discourse and collaborating with each other, autonomously generated knowledge can be expected to explode. The cumulative knowledge of the human race will increase by the combined effort of millions of cogs all over the world. In fact, we foresee an explosion of knowledge, an exponential growth, when cogs begin working with the knowledge generated by other cogs. This kind of cognitive work can proceed without the intervention of a human and therefore proceed at a dramatically accelerated rate. We can easily foresee the point in time where production of new knowledge by cogs exceeds, forever, the production of new knowledge by humans.

In fact, we anticipate a class of *discovery engine cogs* whose sole purpose is to reason about enormous stores of knowledge and continuously generate new knowledge of ever-increasing value resulting ultimately in new discoveries that would have never been discovered by humans or, at the very least, taken humans hundreds if not thousands of years to discover.

12.5 Intellectual Property Ownership

With cognitive systems becoming able to generate knowledge on their own with minimal or no human supervision or interaction, interesting questions as to the ownership of this knowledge arises. In 2019, the Unites States Patent and Trademark Office (USPTO) sought guidance on ownership and intellectual property rights issues stemming from the use of artificial intelligence and cognitive systems to generate knowledge (Deahl, 2019).

The USPTO issued thirteen questions (USPTO, 2019):

1. Should a work produced by an AI algorithm or process, without the involvement of a natural person contributing expression to the resulting work, qualify as a work of authorship protectable under U.S. copyright law?

2. Assuming involvement by a natural person is or should be required, what kind of involvement would or should be sufficient so that the work qualifies for copyright protection? For example, should it be sufficient if a person (i) designed the AI algorithm or process that created the work; (ii) contributed to the design of the algorithm or process; (iii) chose data used by the algorithm for training or otherwise; (iv) caused the AI algorithm or process to be used to yield the work; or (v) engaged in some specific combination of the foregoing activities? Are there other contributions a person could make in a potentially

copyrightable AI-generated work in order to be considered an "author"?

3. To the extent an AI algorithm or process learns its function(s) by ingesting large volumes of copyrighted material, does the existing statutory language (*e.g.*, the fair use doctrine) and related case law adequately address the legality of making such use? Should authors be recognized for this type of use of their works? If so, how?

4. Are current laws for assigning liability for copyright infringement adequate to address a situation in which an AI process creates a work that infringes a copyrighted work?

5. Should an entity or entities other than a natural person, or company to which a natural person assigns a copyrighted work, be able to own the copyright on the AI work? For example: Should a company who trains the artificial intelligence process that creates the work be able to be an owner?

6. Are there other copyright issues that need to be addressed to promote the goals of copyright law in connection with the use of AI?

7. Would the use of AI in trademark searching impact the registrability of trademarks? If so, how?

8. How, if at all, does AI impact trademark law? Is the existing statutory language in the Lanham Act adequate to address the use of AI in the marketplace?

9. How, if at all, does AI impact the need to protect databases and data sets? Are existing laws adequate to protect such data?

10. How, if at all, does AI impact trade secret law? Is the Defend Trade Secrets Act (DTSA), 18 U.S.C. 1836 *et seq.*, adequate to address the use of AI in the marketplace?

11. Do any laws, policies, or practices need to change in order to ensure an appropriate balance between maintaining trade secrets on the one hand and obtaining patents, copyrights, or other forms of intellectual property protection related to AI on the other?

12. Are there any other AI-related issues pertinent to intellectual property rights (other than those related to patent rights) that the USPTO should examine?

13. Are there any relevant policies or practices from intellectual property agencies or legal systems in other countries that may help inform USPTO's policies and practices regarding intellectual property rights (other than those related to patent rights)?

Some think if a programmer develops a system and the system creates new knowledge then the programmer should own the intellectual property rights for that knowledge. However, think about a sculptor using a hammer and chisel to create a masterpiece of art. No one would think the person who built the hammer or chisel should have any claim of ownership of the art. So what is the difference? One difference is the hammer and chisel are tools. They do not create anything on their own. One argument then is cognitive systems like Ashe are mere tools, although sophisticated tools, and so the humans developing or operating the system should have ownership of the products created by the system.

A counter argument is the view cogs are the entities creating the new knowledge, not the programmer. Some feel artificially intelligent systems should own the intellectual property. Still others feel no one owns the intellectual property and instead it is owned by the human race at large. The Levels of Cognitive Augmentation presented in Chapter 3 come into play here. The sculptor using a hammer and chisel represent very low levels of augmentation. Any tool augments human performance. However, the situation becomes less clear as we get to Level 4 and Level 5 of cognitive augmentation where the artificial system is doing most of the cognitive work. Once a system achieves a certain level of autonomy, does it then earn the right to own the intellectual property it produces?

One of the most useful features of cognitive systems is their ability to consume vast amounts of unstructured information and process it much faster than any human ever could. However, in doing so, systems often acquire information via the Internet. Some of the information on the Internet is public but some of it is protected by intellectual property rights such as copyrights. Question #3 from the USPTO asks, and rightly so, what are the legal ramifications of one system using protected material to create new knowledge. Humans do this all of the time. Imagine the situation where a human reads a copyrighted book and learns something he or she then uses to create new knowledge. No one thinks the author of the copyrighted book has any claim to the newly created knowledge. However, it is appropriate for a person to give credit for others' ideas. Are we looking at a future where researchers and scientists must cite autonomous and semi-autonomous cogs?

Most seem to agree if a company owns a cognitive system, then it owns anything the cognitive system creates. One argument here is to equate the cognitive system to any kind of tool or machine on an assembly line. If a worker uses a powered socket wrench to attach parts to a automobile chassis as it moves down the assembly line, no one thinks the socket wrench has any more claim to the automobile than the assembly line worker. The company owns the tools, the raw materials, and employs the worker, so it owns what is produced. However, imagine a cog one day discovers the cure to all cancers, or discovers how to triple the lifetime of every human, something of tremendous value to the entire human race. Should the company own this knowledge? Should it be allowed for everyone in the world be beholden to one company (or one government for that matter) which happens to have stumbled on a discovery of universal importance? Some antitrust laws may speak to this issue, but the issue gets less clear as the value, or perceived value, of something gets enormous. At some point, many feel those kinds of things should belong to everyone. Questions like these and many others will have to be answered by the next generation.

Chapter 13
The Democratization of Expertise

None of the ideas discussed in this book are science fiction. Some may appear to be because they are visions of things not taking place for several years. But all are already at some stage of development. Sweeping changes can occur in a relatively short time. For example, not very long ago, a smartphone could have been called science fiction. However, billions of people across the world are now dependent on the technology and mass-adoption has led to significant cultural, societal, and economic changes. In another example, the first personal computers became available in the mid and late 1970s, but widespread adoption of desktop computers in the business office did not happen until ten years later. By the end of the 1980s, nearly everyone in an office job was using a desktop computer. The transformation took less than a decade.

Today, we are at the beginning of a new era, the cognitive systems revolution. Cognitive system technology today is analogous to smartphone technology 25 years ago, desktop computers in the early 1980s, and automobiles in the early 1900s. The technology is here, but still at an early stage of development and not yet adopted by the masses.

Currently, access to expert-level performance, requires an average person to find and interact with a human expert trained in the art. The expert is usually someone charging a substantial fee. However, the cognitive systems era promises to make it possible for the average person to perform at an expert level in virtually any domain by partnering and collaborating with one or more low-cost cogs. We call the mass adoption of cognitive systems technology the *democratization of expertise*. We foresee cogs being developed for any enterprise for which a human can be an expert. When anyone can be an expert in a field everything changes. Like any technological revolution before, the democratization of expertise will change the way we live, work, play, and evolve.

Earlier this book discussed several existing cognitive systems able to outperform their human counterparts. However, these systems are not geared for the mass market. Instead, they are owned and operated by large companies, government organizations, or are productivity-enhancing applications catering to a professional niche. This is about to change. We will soon see the development of *personal cogs* intended for use by the average person. The range of these cogs will traverse all of our daily lives: teaching, research, companionship, relationship advice, fashion advice, entertainment, cooking, diet coaching, personal training, shopping, etc. Every activity the average person engages in will be touched by cognitive system technology.

World-Wide-Web technology (the markup language, browsers and servers) became available in the early 1990s. Although the Internet had been in existence for 25 years at the time, the Internet was not the mass-market phenomenon as we know it today. Most people in the early 1990s did not use, or even know about, the Internet. Web technology, however, changed the game by making it possible for anyone, with only basic computer knowledge and skill, able to create and host Web pages. This led to the mass adoption of the Internet and the "dot com boom" of the mid to late 1990s. Mass adoption of the Web led to sweeping changes over the next 20 years such as: social media, social networking, online shopping, and streaming media. Many of today's leading companies and services were spawned by the Web revolution: Google, Facebook, Amazon, YouTube, eBay, Twitter, Wikipedia, LinkedIn, Snapchat, Instagram, etc. Online shopping is transforming how we purchase goods and services. Apple's iTunes and YouTube disrupted the music market. Most people today stream musical content rather than purchase physical media such as records, cassettes, and compact disks. Many major shopping store icons have recently closed because of pressure from online shopping. The entertainment industry is in transformation as well with movies and television shows being accessed by streaming services over the Internet rather than via broadcast services. Mass adoption of cognitive systems technology in the cog era will lead to similar disruptions. We foresee the emergence of several new mass-market products and services revolving around cognitive systems technology.

One of the first touchpoints between the masses and cognitive systems technology figures to be voice-controlled assistants like Apple's Siri, Microsoft's Cortana, Google Now, and Amazon Echo's Alexa. Millions already speak to these devices on a daily basis. We expect the voice-controlled entities in these devices to become increasingly intelligent and able to perform higher-level cognitive processing. One of the first milestones will be conversational and contextual natural language dialog. We will soon be able to carry on an extended conversations with these devices. This will bring hundreds of millions of people in natural contact with cognitive systems much like Web technology enabled millions to come in contact with the Internet. Many will not even realize they are engaging with a new type of technology, they will simply know their devices are getting smarter.

The trajectory of systems, devices, and apps getting smarter leads to expert-level performance. What will the world be like when most of us can perform as well as an expert in any field? Are we predicting the end of human experts? Will cogs replace humans? The short answer is no. In fact, we see the cog era creating millions of new jobs just as the computer, Internet, and social media industries have over the last 40 years.

The cog era as we envision it makes expertise available to a vastly larger market than is possible today. For example, today, it is not possible for hundreds of millions of people to be sitting in lawyers' offices on a given day. The prohibitive cost of legal services governs and limits the supply and demand of legal expertise. However, when synthetic lawyers become available to the masses, hundreds of millions of people will be accessing expert legal services daily thus creating a new market impossible today. Humans will be employed in various roles to service this enormous market explosion and the revenues this market will create will be measured in the billions of dollars. We foresee the same phenomenon happen across many domains. Mass adoption of synthetic expertise will create new market opportunities we can only imagine today.

Any technological revolution spawns its own class of entrepreneurs. The cog era and the new market potentials will stimulate the rise of a healthy generation of cog entrepreneurs. What they create is certain to surprise us and also change our market behaviors.

Earlier in this book we explained how cognitive systems learn from unstructured data via unsupervised learning. However, this does not mean cognitive systems will simply spring up out of nowhere on their own. Companies and organizations will produce and train cognitive systems and endow them with special characteristics only that company or organization can provide. Companies and organizations will compete with each other over market share by marketing better cogs than their competitors. Imagine, for example, advertisements for the Goldman Sachs investment cog touting how much better their cog is than the Morgan Stanley cog. We foresee all major entities in every industry entering the cognitive systems industry much like all such entities have embraced the computer and Internet technologies. This will create a demand for a new kind of human expert—the *expertise worker.*

Expertise workers will not only be employed to create highly-capable cognitive systems, but great value will be placed on those employees better able to utilize cogs. Today, it is common for employers to test prospective employees on their computer, Internet, and social media skills. In the cog era, employers will test new hires for their cog skills. The future will belong to those best able to collaborate with cognitive systems and employers will expect this from their workers and will compensate such expertise workers accordingly.

In the age of synthetic expertise, knowledge will be the product of a mixture of biological and artificial thinking. This symbiotic partnership with humans gives rise to an interesting idea. We foresee there will arise the need and opportunity for people to work with cogs and develop their own unique store of knowledge, something we call an expertise knowledge base, or *e-base* for short. We think e-bases could become the next unit of value in human creative endeavor. An e-base will consist of a person's personal cog and all of the intellectual capital it created with the person. Any person will be able to create an e-base. People in industries such as financial services, investment services, legal, medical, news, politics, fishing, cooking, and technology will compete in offering access to their "superior" store of knowledge.

Today, people write books (printed books and e-books) and make YouTube videos to capture their own expertise with a subject matter. But these are static, two-dimensional creations. You can't ask a book or a video questions. Unless the author includes it explicitly, a book or video can't explain how or why something appears nor the chain of events and thinking leading the author to it.

An e-base is different and adds new dimensions to knowledge and expertise. Because the e-base is housed within a cognitive system, it will be able to interact with its users. Interaction will be conversational natural language, visual, auditory, augmented reality, mixed reality, and enhanced reality. In fact, an e-base becomes a virtual representative of the author(s) and the knowledge contained therein with the ability to live forever (Fulbright, 2017a). Imagine a textbook you can query and teach you about what it contains. Imagine a song or artwork by an artist you can not only hear and see, but interact with and live along with them during the creative process itself. Much better than a picture of video, imagine a departed loved one leaving e-bases behind allowing those of us left to not only reminisce but interact with visually, aurally, and cognitively. In the cog era, knowledge and expertise become commodities.

Bibliography

Abràmoff, M. D., Lavin, P. T., Birch, M., Shah, N. and Folk, J. C. 2018. Pivotal trial of an autonomous AI-based diagnostic system for detection of diabetic retinopathy in primary care offices. Digital Med., 1: 39.

Ackoff, R. 1989. From data to wisdom. Journal of Applied Systems Analysis, 16.

Ahaskar, A. 2018. How Gait Analysis is Helping the Police, LiveMint Internet site located at: https://www.livemint.com/Technology/ iEWMTaOcAHJDlSYqbmZRtN/How-gait-analysis-is-helping- the-police. html, last accessed December 2019.

[AI Dreams]. 2013. Steve Worswick Interview—Loebner 2013 winner, AI Dreams Internet site located at: https://aidreams.co.uk/forum/index. php?page=Steve_Worswick_Interview_-_Loebner_2013_winner#. XgDidBdKhUM, last accessed December 2019.

[AISB] The Society for the Study of Artificial Intelligence and Simulation of Behaviour. 2019. AISB X: Creativity Meets Economy (incorporating the Loebner Prize), Available online: https://aisb.org.uk/new_site/?page_id=2, last accessed September 2019.

Al-Emran, M. and Shaalan, K. 2014. A Survey of Intelligent Language Tutoring Systems.

[Alexa Prize]. 2019. Amazon Internet site located at: https://developer.amazon. com/alexaprize, last accessed December 2019.

[Allen Institute]. 2019. Internet site: https://allenai.org/, last accessed November 2019.

Anderson, J., Boyle, C. and Reiser, B. 1985. Intelligent tutoring systems. Science, Vol. 228, Iss. 4698, April.

Anderson, J. R. 2013. The Architecture of Cognition. Psychology Press.

Anderson, L. W., Krathwohl, D. R., Airasian, P. W., Cruikshank, K. A. and Mayer, R. E. 2001. A taxonomy for learning, teaching, and assessing: A revision of Bloom's taxonomy of educational objectives. Pearson.

Anderson, M. and Perrin, A. 2017. Technology use among seniors, Pew Research Center Internet site located at: https://www.pewresearch.org/ internet/2017/05/17/technology-use-among-seniors/, last accessed December 2019.

[AP]. 2019. Self-Help, American Psychological Association Dictionary of Psychology Internet site located at: https://dictionary.apa.org/self-help, last accessed December 2019.

[Apple]. 1987. Knowledge Navigator. Available on the Internet: https://www. youtube.com/watch?v=JIE8xk6Rl1w Last accessed April 2016.

[Apple]. 2015. Siri, Apple Internet page located at: http://www.apple.com/ios/ siri/ and last accessed November 2015.

Ashby, W. R. 1956. An Introduction to Cybernetics. Chapman and Hall, London.

Baker, L. and Hui, F. 2017. Innovations of AlphaGo, DeepMind Blog. Available online: https://deepmind.com/blog/article/innovations-alphago, last accessed November 2019.

Barnes, M., Chen, J. Y. C. and Hill, S. 2017. Humans and Autonomy: Implications of Shared Decision-Making for Military Operations, Army Research Laboratory, retrieved October 6, 2018, from http://www.arl.army.mil/arlreports/2017/ ARLTR-7919.pdf.

Bell, R. 2018. Artificial Intelligence, Automation and the Future of Talent Acquisition, Workforce Internet site located at: https://www.workforce. com/2018/06/15/artificial-intelligence-automation-and-the-future-of-talent-acquisition/, last accessed December 2019.

Berliner, H. 1977. BKG, a Program That Plays Backgammon, Computer Science Department, Carnegie-Mellon University. Available online: https://bkgm. com/articles/Berliner/BKG-AProgramThatPlaysBackgammon/, last accessed November 2019.

Berliner, H. 1980. Backgammon Computer Program Beats World Champion, Artificial Intelligence, 14: 205–220. Available online: https://bkgm.com/ articles/Berliner/BackgammonProgramBeatsWorldChamp/, last accessed November 2019.

Berry, S. K. 2018. A 'smart' toilet could stop us from flushing away our most valuable health information, argues this doctor, CNBC News Internet site located at: https://www.cnbc.com/2018/11/21/smart-toilets-would-be-a-huge-boon-to-public-health-commentary.html, last accessed December 2019.

Bhagwat, V. A. 2018. Deep Learning for Chatbots. San Jose State University.

Birch, I., Gwinnett, C. and Walker, J. 2016. Aiding the interpretation of forensic gait analysis: Development of a features of gait database. Science and Justice. Available online at: https://core.ac.uk/download/pdf/46665554.pdf, last accessed December 2019.

Bishop, C. 2006. Pattern Recognition and Machine Learning. Springer.

[BlackBoiler]. 2019. BlackBoiler Issued Patents from USPTO for AI-Assisted Contract Review Tool, BlackBoiler Internet site: https://www.blackboiler. com/post/blackboiler-issues-patents, last accessed November 2019.

Bloom, B. S., Engelhart, M. D., Furst, E. J., Hill, W. H. and Krathwohl, D. R. 1956. Taxonomy of educational objectives: The classification of educational goals. Handbook I: Cognitive Domain. New York: David McKay Company.

Boiano, S., Borda, A., Gaia, G., Rossi, S. and Cuomo, P. 2018. Chatbots and New Audience Opportunities for Museums and Heritage Organisations, in Electronic Visualisation and the Arts (EVA).

Breazeal, C. 2000. Sociable Machines: Expressive Social Exchange Between Humans and Robots, Massachusetts Institute of Technology. Available online: http:// groups.csail.mit.edu/lbr/hrg/2000/phd.pdf, last accessed November 2019.

Brem, A., Rauschnabel, P. and Ro, Y. 2015. Augmented Reality Smart Glasses: Definition, Conceptual Insights, and Managerial Importance. Available online: https://tinyurl.com/y4gal28o, last accessed December 2019.

Bright, P. 2016. Tay, the neo-Nazi millennial chatbot, gets autopsied, Ars Technica, 25 March 2016. Available online: https://arstechnica.com/information-technology/2016/03/tay-the-neo-nazi-millennial-chatbot-gets-autopsied/, last accessed October 2019.

Brooks, R. 1986. A robust layered control system for a mobile robot. IEEE Journal of Robotics and Automation, 2(1): 14–23.

Buchanan, B. G. and Shortliffe, E. 1980. Rule-Based Expert Systems: The MYCIN Experiments of the Stanford Heuristic Programming Project, Addison-Wesley. Available online: http://people.dbmi.columbia.edu/~ehs7001/Buchanan-Shortliffe-1984/MYCIN%20Book.htm last accessed November 2019.

Buitron, C. 2017. Rise of the $300 Billion Senior Care Industry, Huffington Post Internet site located at: https://www.huffpost.com/entry/rise-of-the-300-billion-senior-care-industry_b_58b0c35ce4b0e5fdf61971ee, last accessed December 2019.

Burns, M. 2019. BMW's magical gesture control finally makes sense as touchscreens take over cars, TechCrunch Internet site located at: https://techcrunch.com/2019/11/04/bmws-magical-gesture-control-finally-makes-sense-as-touchscreens-take-over-cars/, last access December 2019.

Bush, N. and Wallace, R. 2001. AIML 1.0 Standard Passed: Free ("Open Source") Artificial Intelligence Markup Language Brings Bots to the Masses, A.L.I.C.E. AI Foundation Internet site. Located at: https://web.archive.org/web/20070715113602/http://www.alicebot.org/press_releases/2001/aiml10.html, last accessed December 2019.

Bush, V. 1945. As We May Think. The Atlantic, July.

Buskirk, E. 2009. BellKor's Pragmatic Chaos Wins $1 Million Netflix Prize by Mere Minutes, Wired. Available online: https://www.wired.com/2009/09/bellkors-pragmatic-chaos-wins-1-million-netflix-prize/, last accessed November 2019.

Carbonell, J. R. 1970. AI in CAI: An artificial-intelligence approach to computer-assisted instruction. IEEE Transactions on Man-Machine Systems, Vol. 11, Iss. 4, December.

Carbonell, J. R. 1983. Machine learning: A historical and methodological analysis. AI Magazine, 4(3): 69–79.

[Catalia Health]. 2019. How Mabu Works, Catalia Health Internet site located at http://www.cataliahealth.com/how-it-works/, last accessed December 2019. Video available at: https://www.youtube.com/watch?v=A3XwzlvOW7k), last accessed December 2019.

Caudell, T. and Mitzell, D. 1992. Augmented Reality: An Application of Heads-Up Display Technology to Manual Manufacturing Processes, Proceedings of the Twenty-Fifth Hawaii International Conference on Systems Sciences, Hawaii. Available online: https://tinyurl.com/yauems8b, last accessed December 2019.

Chakrabarti, S., Ester, M., Fayyad, U., Gehrke, J., Han, J., Morishita, S., Piatetsky-Shapiro, G. and Wang, W. 2006. Data mining curriculum: A proposal (Version 0.91). Intensive Working Group of ACM SIGKDD Curriculum Committee. Available online: http://www.sigkdd.org/curriculum/CURMay06.pdf, last accessed November 2019.

Chandola, V., Sukumar, S. R. and Schryver, J. 2013. Knowledge discovery from massive healthcare claims data. KDD, pp. 1312–1320, Chicago.

Chapman, S. 1942. Blaise Pascal (1623–1662) Tercentenary of the calculating machine. Nature, London 150: 508–509.

Chase, W. and Simon, H. 1973. Perception in chess. Cognitive Psychology, Volume 4.

Chen, J. Y. C., Stowers, K., Barnes, M. J., Selkowitz, A. R. and Lakhamni, S. G. 2017. Human-autonomy teaming and agent transparency. HRI '17 Proceedings of the Companion of the 2017 ACM/IEEE International Conference on Human-Robot Interaction, Association for Computing Machinery, March 2017.

Chen, X., Shrivastava, A. and Gupta, A. 2013. NEIL: Extracting visual knowledge from web data. Proceedings of the 2013 IEEE International Conference on Computer Vision, IEEE.

[ChessBase]. 2018. AlphaZero: Comparing Orangutans and Apples, ChessBase Internet page: https://en.chessbase.com/post/alpha-zero-comparing-orang-utans-and-apples, last accessed February 2018.

Christensen, Clayton M. 1997. The innovator's dilemma: when new technologies cause great firms to fail, Boston. Massachusetts, USA: Harvard Business School Press, ISBN 978-0-87584-585-2.

Chu, J. and Wang, C. 2019. Meet your new colleague—artificial intelligence, Deloitte Internet site: https://www2.deloitte.com/cn/en/pages/technology/articles/meet-your-new-colleague-artificial-intelligence.html, last accessed November 2019.

Clancey, W. J. 1986. Intelligent Tutoring Systems: A Tutorial Survey, September, Stanford University report: STAN-CS-87-1174 (KSL-86-58). Available online: https://apps.dtic.mil/dtic/tr/fulltext/u2/a187066.pdf, last accessed November 2019.

Coffield, F., Moseley, D., Hall, E. and Ecclestone, E. 2004. Learning styles and pedagogy in post-16 learning: a systematic and critical review, London: Learning and Skills Research Centre. Available online: https://web.archive.org/web/20160304072804/http://sxills.nl/lerenlerennu/bronnen/Learning%20styles%20by%20Coffield%20e.a..pdf, last accessed November 2019.

Colby, K. M., Hilf, F. D., Weber, S. and Kraember, H. 1972. Turing-like indistinguishability tests for the validation of a computer simulation of paranoid processes. Artificial Intelligence, 3(C): 199–221.

Colon, A. and Greenwald, M. 2015. Amazon Echo, PC Magazine Internet page located at: http://www.pcmag.com/article2/0,2817,2476678,00.asp, last accessed November 2015.

Comstock, J. 2018. Inui Health, formerly Scanadu, announces FDA-cleared home urine testing platform, Mobile Health News Internet site located at: https://www.mobihealthnews.com/content/inui-health-formerly-scanadu-announces-fda-cleared-home-urine-testing-platform, last accessed December 2019.

Conway, B. 2019. The 7 Learning Styles: What's Your Learning Style?, Employee Connect Internet site: https://www.employeeconnect.com/blog/seven-7-learning-styles/, last accessed November 2019.

Corbett, A. T. and Anderson, J. R. 1992. LISP intelligent tutoring system research in skill acquisition. *In*: Larkin, J. and Chabay, R. (eds.). Computer Assisted Instruction and Intelligent Tutoring Systems: Shared Goals and Complementary Approaches. Englewood Cliffs, New Jersey: Prentice-Hall Inc.

Cortes, C. and Vapnik, V. N. 1995. Support-vector networks. Machine Learning, 20(3): 273–297.

Crevier, D. 1993. AI: The Tumultuous Search for Artificial Intelligence. New York, NY: BasicBooks.

[DARPA]. 2019. The Grand Challenge, DARPA Internet site: https://www.darpa.mil/about-us/timeline/-grand-challenge-for-autonomous-vehicles, last accessed November 2019.

[DARPA]. 2019b. The Robotics Challenge, DARPA Internet site: https://www.darpa.mil/program/darpa-robotics-challenge, last accessed November 2019.

Deahl, D. 2019. The USPTO wants to know if artificial intelligence can own the content it creates, The Verge, November 13, 2019. Available online: https://www.theverge.com/2019/11/13/20961788/us-government-ai-copyright-patent-trademark-office-notice-artificial-intelligence, last accessed November 2019.

Dechter, R. 1986. Learning while searching in constraint-satisfaction problems. AAAI-86 Proceedings. Available online: https://www.aaai.org/Papers/AAAI/1986/AAAI86-029.pdf, last accessed November 2019.

[DeepMind]. 2018a. The story of AlphaGo so far, DeepMind Internet page: https://deepmind.com/research/alphago/, last accessed February 2018.

[DeepMind]. 2018b. AlphaGo Zero: learning from scratch, DeepMind Internet page: https://deepmind.com/blog/alphago-zero-learning-scratch/, last accessed February 2018.

de Groot, A. D. 1965. Thought and Choice in Chess. The Hague, Mouton.

DeJong, G. 1981. Generalizations based on explanations. pp. 67–69. *In*: IJCAI'81, The Seventh International Joint Conference on Artificial Intelligence. Vancover, BC.

DeJong, G. and Lim, S. 2017. Explanation-based learning. *In*: Sammut, C. and Webb, G. I. (eds.). Encyclopedia of Machine Learning and Data Mining. Springer, Boston, MA.

Deng, L. and Yu, D. 2014. Deep learning: Methods and applications. Foundations and Trends in Signal Processing, 7: 3–4.

Deshpande, A., Shahane, A., Gadre, D., Deshpande, M. and Joshi, P. 2017. A survey of various chatbot implementation techniques. International Journal of Computer Engineering and Applications, Volume XI, Special Issue, May 17.

Dickmanns, E. D. 2007. Dynamic Vision for Perception and Control of Motion. Springer Verlag, London.

[Dinsow]. 2019. CT Asia Robotics Internet site located at https://www.dinsow.com, last accessed December 2019.

Donovan, T. 2010. Replay: The History of Video Games, Yellow Ant.

Dreyfus, H. L. 1972. What Computers Can't Do: A Critique of Artificial Reason. The MIT Press, Cambridge, MA.

Dreyfus, H. L. and Dreyfus, S. E. 1988. Mind Over Machine: The Power of Human Intuition and Expertise in the Era of the Computer. New York: Free Press.

Drucker, P. F. 2006. Innovation and Entrepreneurship. Harper Business.

[Economist]. 2016. From not working to neural networking, The Economist, June 25. Available online: https://www.economist.com/special-report/2016/06/23/from-not-working-to-neural-networking, last accessed November 2019.

Eisenstein, E. L. 1983. The Printing Revolution in Early Modern Europe. Cambridge University Press.

Ekman, P., Freisen, W. and Ancoli, S. 1980. Facial designs of emotional experience. Journal of Personality and Social Psychology, Vol. 39.

Ekman, P. 2019. Facial Action Coding System (FACS), Ekman Group Internet site located at: https://www.paulekman.com/facial-action-coding-system/, last accessed December 2019.

[ElliQ]. 2019. Hi, I'm ElliQ, Intuition Robotics Internet site located at: https://elliq.com, last accessed December 2019.

Engelbart, D. C. 1962. Augmenting Human Intellect: A Conceptual Framework, Summary Report AFOSR-3233, Stanford Research Institute, Menlo Park, CA, October 1962.

Esfandiari, N. 2014. Knowledge discovery in medicine: Current issue and future trend. Science Direct, 41(9): 4434–4463, July.

Fayyad, U., Piatetsky-Shapiro, G. and Smyth, P. 1996. From data mining to knowledge discovery in databases. AI Magazine, Vol. 17, No. 3. Available online: https://www.kdnuggets.com/gpspubs/aimag-kdd-overview-1996-Fayyad.pdf, last accessed November 2019.

Feigenbaum, E. A. and Simon, H. A. 1984. EPAM-like models of recognition and learning. Cognitive Science, 8: 305–336.

Feldman, S. and Reynolds, H. 2014. Cognitive computing: A definition and some thoughts. KM World, Vol. 23, Issue 10.

Feldman, S. 2016. What is Cognitive Computing? KM World Internet page located at http://www.kmworld.com/Articles/Editorial/ViewPoints/What-is-Cognitive-Computing-108931.aspx and last retrieved March 2016.

Ferrucci, D., Brown, E., Chu-Carroll, J., Fan, J., Gondek, D., Kalyanpur, A., Lally, A., Murdock, J. W., Nyberg, E., Prager, J., Schlaefer, N. and Welty, C. 2010. Building Watson: An overview of the DeepQA project. AI Magazine, Vol. 31, No. 3.

Ferrucci, D. A. 2012. Introduction to "This is Watson." IBM J. Res. & Dev. Vol. 56, No. 3/4.

[Field Agent]. 2015. Millennials, Boomers, & 2015 Resolutions: 5 Key Generational Differences, Field Agent Internet site located at: https://blog.fieldagent.net/millennials-boomers-new-years-resolutions-5-key-generational-differences, last accessed December 2019.

Fisher, J. 2019. The best 360 Cameras for 2019, PC Mag. Available at: https://www.pcmag.com/roundup/354276/the-best-360-cameras, last accessed December 2019.

Fleming, N. and Baume, D. 2006. Learning styles again: VARKing up the right tree! Educational Developments, 7(4): 4–7.

Forbus, K. and Hinrichs, T. 2006. Companion cognitive systems: A step toward human-level AI. AI Magazine, Vol. 27, No. 2.

Frawley, W. J. 1992. Knowledge discovery in databases: An overview. AI Magazine, Vol. 13, No. 3.

Freedman, R., Ali, S. and Mcroy, S. 2000. What is an intelligent tutoring system? Intelligence, Vol. 11, Iss. 3.

Friedman, T. L. 1999. The Lexus and the Olive Tree: Understanding Globalization. New York: Random House.

Fukushima, K. 1980. Neocognitron: A self-organizing neural network model for a mechanism of pattern recognition unaffected by shift in position. Biological Cybernetics, 36[4]: 193–202.

Fulbright, R. 2002. Information Domain Modeling of Emergent Systems, Technical Report CSCE 2002–014, May 2002, Department of Computer Science and Engineering, University of South Carolina, Columbia, SC.

Fulbright, R. 2016a. The cogs are coming: the coming revolution of cognitive computing. Proceedings of the 2016 Association of Small Computer Users in Education (ASCUE) Conference, June.

Fulbright, R. 2016b. How personal cognitive augmentation will lead to the democratization of expertise. Fourth Annual Conference on Advances in Cognitive Systems, Evanston, IL, June 2016. Available at http://www.cogsys.org/posters/2016, last retrieved January 2017.

Fulbright, R. 2017a. ASCUE 2067: How we will attend posthumously. Proceedings of the 2017 Association of Small Computer Users in Education (ASCUE) Conference, June 2017.

Fulbright, R. 2017b. Cognitive augmentation metrics using representational information theory. *In*: Schmorrow, D. and Fidopiastis, C. (eds.). Augmented Cognition. Enhancing Cognition and Behavior in Complex Human Environments. AC 2017. Lecture Notes in Computer Science, Vol. 10285. Springer.

Fulbright, R. 2018. On measuring cognition and cognitive augmentation. *In*: Yamamoto, S. and Mori, H. (eds.). Human Interface and the Management of Information. LNCS 10904. Proceedings of HCI International 2018 Conference. Las Vegas, NV, Springer.

Fulbright, R. 2019. Calculating cognitive augmentation—A case study. *In*: Schmorrow, D. and Fidopiastis, C. (eds.). Augmented Cognition. HCII 2019. Lecture Notes in Computer Science, Vol. 11580. Springer, Cham.

Fulbright, R. 2020. The expertise level. Proceedings of the HCI International 2020 Conference, Springer, Copenhagen, In press.

Fulbright, R. and Walters, G. 2020. Synthetic expertise. Proceedings of the HCI International 2020 Conference, Springer, Copenhagen, In press.

Gardner, H. 2011. Frames of Mind. 3rd edition, Basic Books.

[Gartner]. 2019. Gartner Hype Cycle, Gartner Internet site located at: https://www.gartner.com/en/research/methodologies/gartner-hype-cycle,   last accessed December 2019.

Gaskin, J. 2008. Whatever happened to artificial intelligence? Computerworld, June 24. Available online: https://www.computerworld.com/article/2534413/what-ever-happened-to-artificial-intelligence-.html##targetText=In%20 1965%2C%20artificial%20intelligence%20innovator,work%20a%20man%20 can%20do.%22&targetText=The%20%E2%80%9CAs%20a%20Service%E2-

%80%9D%20model,that%20align%20to%20business%20outcomes, last accessed November 2019.

Genesereth, M. and Nilsson, N. 1987. Logical Foundations of Artificial Intelligence. Morgan Kaufmann.

Gil, D. 2014. Cognitive systems and the future of expertise, YouTube video located at https://www.youtube.com/watch?v=0heqP8d6vtQ and last accessed May 2019.

Gil, Y., Garijo, D., Ratnakar, V., Mayani, R., Adusumilli, R., Boyce, H. and Mallick, P. 2016. Automated hypothesis testing with large scientific data repositories. In Proceedings of the Fourth Annual Conference on Advances in Cognitive Systems (ACS), Evanston, IL.

Gobet, F. and Simon, H. 2000. Five Seconds or Sixty? Presentation time in expert memory. Cognitive Science, Vol. 24, No. 4.

Gobet, F., Lane, P. C. R., Croker, S., Cheng, P. C. H., Jones, G., Oliver, I. and Pine, J. M. 2001. Chunking mechanisms in human learning. Trends in Cognitive Sciences, 5: 236–243.

Gobet, F. and Chassy, P. 2009. Expertise and intuition: a tale of three theories. Minds & Machines, Springer.

Gobet, F. 2016. Understanding Expertise: A Multidisciplinary Approach. Palgrave, UK.

Goel, A. K. and Polepeddi, L. 2016. Jill Watson: A Virtual Teaching Assistant for Online Education Design & Intelligence Laboratory, School of Interactive Computing, Georgia Institute of Technology. Available online: https://smartech.gatech.edu/bitstream/handle/1853/59104/goelpolepeddi-harvardvolume-v7.1.pdf, last accessed November 2019.

Goldhammer, F., Moosbrugger, H. and Kawietz, S. 2009. FACT-2—The Frankfurt adaptive concentration test: Convergent validity with self-reported cognitive failures. European Journal of Psychological Assessment, Vol. 25, No. 2, January.

Goode, L. 2018. Touchless Gesture Controls on Phones? Think Bigger, Wired Internet site located at: https://www.wired.com/story/gesture-controls-phones-samsung-lg-google/, last accessed December 2019.

[Google]. 2015. Google Now: What is it? Google Internet page located at: https://www.google.com/landing/now/#whatisit, last accessed November 2015.

Gregory, M. 2019. AI Trained on Old Scientific Papers Makes Discoveries Humans Missed, Vice Internet page located at: https://www.vice.com/en_in/article/neagpb/ai-trained-on-old-scientific-papers-makes-discoveries-humans-missed and last accessed August 2019.

Grobelnik, M. and Mladenić, D. 2005. Automated knowledge discovery in advanced knowledge management. Journal of Knowledge Management, 9(5): 132–149.

Gugliocciello, G. and Doda, G. 2016. IBM Watson Ecosystem Opens for Business in India, IBM News Release available at https://www-03.ibm.com/press/us/en/pressrelease/48949.wss and last retrieved March 2016.

Güzeldere, G. and Franchi, S. 1995. Dialogues with colorful personalities of early AI. Stanford Humanities Review, Vol. 4, No. 2.

Haenssle, H. A., Fink, C., Schneiderbauer, R., Toberer, F., Buhl, T., Blum, A., Kalloo, A., Hassen, A. B. H., Thomas, L., Enk, A. and Uhlmann, L. 2018. Man against machine: diagnostic performance of a deep learning convolutional neural network for dermoscopic melanoma recognition in comparison to 58 dermatologists. Annals of Oncology, 29(8): 1836–1842. August. Available online: https://academic.oup.com/annonc/article/29/8/1836/5004443, last accessed November 2019.

Hancock, M., Stiers, J., Higgins, T., Swarr, F., Shrider, M. and Sood, S. 2019. A hierarchical characterization of knowledge for cognition. *In*: Schmorrow, D. and Fidopiastis, C. (eds.). Augmented Cognition. HCII 2019. Lecture Notes in Computer Science, Vol. 11580. Springer, Cham.

Hartley, R. V. L. 1928. Transmission of information. Bell System Technical Journal, 7: 535–563.

Hartley, J. R. and Sleeman, D. H. 1973. Towards more intelligent teaching systems. International Study of Man-machine Learning Studies, Vol. 5, Iss. 2.

Haswell, H. and Pelkey, D. 2016. Under Armour and IBM To Transform Personal Health and Fitness. Powered By IBM Watson: New Cognitive Coaching System Will Apply Machine Learning to the World's Largest Digital Health and Fitness Community, IBM News Release available at https://www-03. ibm.com/press/us/en/pressrelease/48764.wss and last retrieved March 2016.

Hayes-Roth, F., Waterman, D. and Lenat, D. 1983. Building Expert Systems. Addison-Wesley.

Havasi, C., Speer, R. and Alonso, J. 2007. ConceptNet 3: A flexible, multilingual semantic network for common sense knowledge. Proceedings of Recent Advances in Natural Language Processing.

Hawkins, J. and George, D. 2006. Hierarchical Temporal Memory—Concepts, Theory, and Terminology. PDF available at http://numenta.com/learn/hierarchical-temporal-memory-white-paper.html, last accessed February 2016.

Hebb, D. O. 1949. The Organization of Behavior. Psychology Press.

Hewitt, C., Bishop, P. and Steiger, R. 1973. A Universal Modular Actor Formalism for Artificial Intelligence. WorryDream, retrieved October 9, 2018, from http://worrydream.com/refs/Hewitt-ActorModel.pdf.

Hewitt, C. 1977. Viewing control structures as patterns of passing messages. Journal of Artificial Intelligence, June.

Ho, T. K. 1995. Random decision forests. Proceedings of the Third International Conference on Document Analysis and Recognition. Montreal, Quebec: IEEE, 1: 278–282. doi:10.1109/ICDAR.1995.598994. ISBN 0-8186-7128-9. Retrieved 5 June 2016.

Hochreiter, S. and Schmidhuber, J. 1997. Long short-term memory. Neural Computation, 9(8): 1735–1780.

Holland, J. H. 1975. Adaptation in Natural and Artificial Systems. The University of Michigan Press, Ann Arbor, MI.

Hooper, R. 2012. Ada Lovelace: My brain is more than merely mortal, New Scientist, Internet page located at https://www.newscientist.com/article/dn22385-ada-lovelace-my-brain-is-more-than-merely-mortal, last accessed November 2015.

Hopfield, J. 1982. Neural networks and physical systems with emergent collective computational abilities. Proceedings of the National Academy of Sciences of the USA, 79(8): 2554–2558.

Huhns, M. N. and Singh, M. P. 1994. Distributed artificial intelligence for information systems. CKBS-94 Tutorial, June 15, University of Keele, UK.

[IBM]. 2014. IBM Forms New Watson Group to Meet Growing Demand for Cognitive Innovations, 2014. IBM Internet page located at: https://www03. ibm.com/press/us/en/pressrelease/42867.wss, last accessed May 2015.

[IBM]. 2015a. IBM Launches Industry's First Consulting Practice Dedicated to Cognitive Business, 2015. IBM Internet page: https://www-03.ibm.com/ press/us/en/pressrelease/47785.wss, last accessed November 2015.

[IBM]. 2015b. Watson Health, IBM Internet page located at: http://www.ibm. com/smarterplanet/us/en/ibmwatson/health/, last accessed November 2015.

[IBM]. 2018. Deep Blue, IBM Internet page: http://www-03.ibm.com/ibm/ history/ibm100/us/en/icons/deepblue/, last accessed February 2018.

[Inventions World]. 2018. Best 5 Personal Healthcare Robots You'll Intend to Buy Soon, YouTube Internet site located at: https://www.youtube.com/ watch?v=cdFpyp1nhCM, last accessed December 2019.

Ireland, C. 2012. Alan Turing at 100, Harvard Gazette, September 2012. Online: https://news.harvard.edu/gazette/story/2012/09/alan-turing-at-100/, last accessed September 2019.

Isaacson, W. 2014. The Innovators: How a Group of Hackers, Geniuses, and Geeks Created the Digital Revolution. Simon & Schuster, New York, NY.

Itamar, A., Rose, D. and Karnowski, P. 2013. Deep machine learning—A new frontier in artificial intelligence research—a survey paper. IEEE Computational Intelligence Magazine.

Jackson, J. 2011. IBM Watson Vanquishes Human Jeopardy Foes, PC World. Internet page http://www.pcworld.com/article/219893/ibm_watson_vanquishes_ human_jeopardy_foes.html, last accessed May 2015.

Jain, A. 2016. The 5 V's of Big Data, IBM Internet site located at: https://www.ibm. com/blogs/watson-health/the-5-vs-of-big-data/, last accessed December 2019.

Jancer, M. 2016. IBM's Watson Takes On Yet Another Job, as a Weather Forecaster, Smithsonian, 26 August 2016. Available online: https://www. smithsonianmag.com/innovation/ibms-watson-takes-yet-another-job-weather-forecaster-180960264/?no-ist, last accessed September 2019.

Jochem, T. M., Pomerleau, D. A. and Thorpe, C. E. 1995. Vision-based neural network road and intersection detection and traversal. IEEE Conference on Intelligent Robots and Systems, August 5–9, 1995, Pittsburgh, Pennsylvania, USA.

Kahney, L. 2018. Your smartphone is ready to take augmented reality mainstream, Wired Magazine Internet site located at: https://www.wired.co.uk/article/ augmented-reality-breakthrough-2018, last accessed December 2019.

Kasparov, G. 2017. Deep Thinking. Public Affairs.

Katz, M. 2017. Welcome to the Era of the AI Coworker. Wired Internet site located at: https://www.wired.com/story/welcome-to-the-era-of-the-ai-coworker/, last accessed December 2019.

Kelly, J. E. and Hamm, S. 2013. Smart Machines: IBMs Watson and the Era of Cognitive Computing. Columbia Business School Publishing. Columbia University Press, New York, NY.

Kerr, I. R. 2005. Bots, Babes and the Californication of Commerce. University of Ottowa.

Kidd, C. 2015. Introducing the Mabu Personal Healthcare Companion. Catalia Health Internet site located at: https://www.cataliahealth.com/introducing-the-mabu-personal-healthcare-companion/, last accessed December 2019.

Kieras, D. E. and Meyer, D. E. 1997. An overview of the EPIC architecture for cognition and performance with application to human-computer interaction. Human-Computer Interaction, 12, Lawrence Erlbaum Associates, Inc.

King, H. 2015. Google files patent for creepy teddy bear, CNN Business Internet site located at: https://money.cnn.com/2015/05/22/technology/google-doll-toy-connected-device-patent/, last accessed December 2019. Patent application US 2015/0138333 available online: https://patentimages.storage.googleapis.com/b9/a8/64/fdcdd6050d083a/US20150138333A1.pdf, last accessed December 2019.

King, J. A. 1995. Intelligent agents: Bringing good things to life. AI Expert, February, 17–19.

Kolb, D. A. 1984. Experiential Learning: Experience as the Source of Learning and Development (Vol. 1). Englewood Cliffs, NJ: Prentice-Hall.

Krizhevsky, A., Sutskever, I. and Hinton, G. E. 2017. ImageNet classification with deep convolutional neural networks. Communications of the ACM, 60(6): 84–90.

Krueger, M. W., Gionfriddo, T. and Hinrichsen, K. 1985. VIDEOPLACE—an artificial reality. ACM SIGCHI Bulletin, 16(4): 35–40.

Kurgan, L. A. 2006. A survey of knowledge discovery and data mining process models. The Knowledge Engineering Review, 21(1): 1–24.

Kurtzman, L. 2019. AI Rivals Expert Radiologists at Detecting Brain Hemorrhages, UCSF Research, Internet site: https://www.ucsf.edu/news/2019/10/415681/ai-rivals-expert-radiologists-detecting-brain-hemorrhages, last accessed November 2019.

Lacoma, T. 2019. The best smart pillows with sleep tracking. Digital Trends Internet site located at: https://www.digitaltrends.com/home/best-smart-pillows-with-sleep-tracking-2/, last accessed December 2019.

Laird, J. E., Rosenbloom, P. S. and Newell, A. 1986. Chunking in soar: The anatomy of a general learning mechanism. Machine Learning, 1(1): 11–46.

Laird, J. E. 2012. The Soar Cognitive Architecture. MIT Press.

Laird, J. E., Lebiere, C. and Rosenbloom, P. S. 2017. A standard model of the mind: toward a common computational framework across artificial intelligence, cognitive science, neuroscience, and robotics. AI Magazine, Vol. 38, No. 4.

Laird, R. and Newell, A. 1987. Soar: An architecture for general intelligence. Artificial Intelligence, 33: 1–64.

Langley, P. and Choi, D. 2006. A unified cognitive architecture for physical agents. Proceedings of the The Twenty-First National Conference on Artificial

Intelligence and the Eighteenth Innovative Applications of Artificial Intelligence Conference, Boston.

Langley, P. 2013. Three challenges for research on integrated cognitive systems. Proceedings of the Second Annual Conference on Advances in Cognitive Systems.

Larson, S. 2016. Microsoft's AI is already talking to millions, The Daily Dot Internet site: https://www.dailydot.com/debug/microsoft-chat-bot-china/ and last accessed February 2020.

Le, Q. V., Ranzato, M., Monga, R., Devin, M., Chen, K., Corrado, G. S., Dean, J. and Ng, A. Y. 2012. Building high-level features using large scale unsupervised learning. arXiv: 1112.6209v5. Available online: https://arxiv. org/pdf/1112.6209.pdf, last accessed November 2019.

Lenat, D. B. and Guha, R. V. 1989. Building Large Knowledge-Based Systems; Representation and Inference in the Cyc Project (1st ed.). Boston, MA, USA: Addison-Wesley Longman Publishing Co., Inc.

Leopold, T. 2017. A professor built an AI teaching assistant for his courses—and it could shape the future of education. Business Insider, 22 March 2017. Available online: https://www.businessinsider.com/a-professor-built-an-ai-teaching-assistant-for-his-courses-and-it-could-shape-the-future-of-education-2017-3, last accessed September 2019.

Lexico. 2019. Oxford University Press, Lexico.com Internet site located at: https://www.lexico.com/en/definition/gesture, last accessed December 2019.

Licklider, J. C. R. 1960. Man-computer symbiosis. IRE Transactions on Human Factors in Electronics, Vol. HFE-1, March.

Lindsay, R. K., Buchanan, B. G., Feigenbaum, E. A. and Lederberg, J. 1980. Applications of Artificial Intelligence for Organic Chemistry: The Dendral Project. McGraw-Hill Book Company.

Linn, A. 2015. Microsoft researchers win ImageNet computer vision challenge, Microsoft Internet site. Available online: https://blogs.microsoft.com/ai/microsoft-researchers-win-imagenet-computer-vision-challenge/, last accessed November 2019.

Linnainmaa, S. 1976. Taylor expansion of the accumulated rounding error. BIT Numerical Mathematics, 16(2): 146–160.

Lloyd, S. 2000. Ultimate physical limits to computation. Nature, 406: 1047–1054.

Longoni, C. and Morewedge, C. K. 2019. AI Can Outperform Doctors. So Why Don't Patients Trust It? Harvard Business Review, October 30. Available online: https://hbr.org/2019/10/ai-can-outperform-doctors-so-why-dont-patients-trust-it, last accessed November 2019.

López, G., Quesada, L. and Guerrero, L. A. 2017. Alexa vs. Siri vs. Cortana vs. Google Assistant: A comparison of speech-based natural user interfaces. In International Conference on Applied Human Factors and Ergonomics.

Markoff, J. 2008. Joseph Weizenbaum, Famed Programmer, Is Dead at 85, The New York Times, 13 March 2008.

Markoff, J. 2011. Computer Wins on Jeopardy!: Trivial, It's Not, The New York Times, 16 February 2011. Available online: https://www.nytimes.com/2011/02/17/science/17jeopardy-watson.html?auth=login-smartlock, last accessed September 2019.

[Mayo Clinic]. 2019. Dementia, Mayo Clinic Internet site located at: https://www.mayoclinic.org/diseases-conditions/dementia/symptoms-causes/syc-20352013, last accessed December 2019.

McCarthy, J., Minsky, M., Rochester, N. and Shannon, C. 1955. A Proposal for the Dartmouth Summer Research Project on Artificial Intelligence. Stanford University Internet page locate at http://www-formal.stanford.edu/jmc/history/dartmouth/dartmouth.html, last accessed November 2015.

McCarthy, J. 1960. Recursive functions of symbolic expressions and their computation by machine. Part I, Communications of the ACM, Vol. 3, Issue 4, April.

McCulloch, W. S. and Pitts, W. 1943. A logical calculus of the ideas immanent in nervous activity. Bulletin of Mathematical Biophysics, 5: 115.

Medler, D. A. 1998. A brief history of connectionism. Neural Computing Surveys, 1: 61–101.

[Mentalup]. 2019. Concentration Test. Mentalup Internet site located at: https://www.mentalup.co/blog/concentration-test, last accessed December 2019.

Michie, D. 1963. Experiments on the mechanization of game-learning. The Computer Journal, Volume 6, Issue 3, November. Available online: http://people.csail.mit.edu/brooks/idocs/matchbox.pdf, last accessed November 2019.

[Microsoft]. 2015. What is Cortana? Microsoft Internet page: http://windows.microsoft.com/en-us/windows-10/getstarted-what-is-cortana, last accessed November 2015.

Milgram, P. and Kishino, A. F. 1994. Taxonomy of mixed reality visual displays. IEICE Transactions on Information and Systems, pp. 1321–1329. Available online: http://etclab.mie.utoronto.ca/people/paul_dir/IEICE94/ieice.html, last accessed December 2019.

Miller, G. A. 1956. The magical number seven, plus or minus two: some limits on our capacity for processing information. Psychological Review, 63: 81–97.

Minsky, M. 1977. Frame-system theory. In: Johnson-Laird, P. N. and Watson, P. C. (eds.). Thinking. Readings in Cognitive Science. Cambridge University Press.

Minsky, M. 1986. The Society of Mind. New York: Simon & Schuster.

Minsky, M. 2007. The Emotion Machine: Commonsense Thinking, Artificial Intelligence, and the Future of the Human Mind. Simon & Schuster.

[MIRROR]. 2019. The MIRROR Internet site located at: https://www.mirror.co/, last accessed December 2019.

Morais, B. 2013. Can Humans Fall in Love with Bots? The New Yorker, 19 November. Available online https://www.newyorker.com/tech/annals-of-technology/can-humans-fall-in-love-with-bots, last accessed September 2019.

Moravec, H. P. 1990. The stanford cart and the CMU rover. In: Cox, I. J. and Wilfong, G. T. (eds.). Autonomous Robot Vehicles. Springer, New York, NY.

Moscaritolo, A. 2017. H&R block enlists IBM Watson to find tax deductions. PC Magazine, 2 February 2017. Available online: https://www.pcmag.com/news/351508/h-r-block-enlists-ibm-watson-to-find-tax-deductions, last accessed September 2019.

Moses, J. 2008. Macsyma: A personal history. Milestones in Computer Algebra.

Moses, J. 2012. Macsyma: A personal history. Journal of Symbolic Computation, 47: 123–130.

Myers, K. D., Knowles, J. W., Staszak, D., Shapiro, M. D., Howard, W., Yadava, M., Zuzick, D., Williamson, L., Shah, N. H., Banda, J. M., Leader, J., Cromwell, W. C., Trautman, E., Murray, M. F., Baum, S. J., Myers, S., Gidding, S. S., Wilemon, K. and Rader, D. J. 2019. Precision screening for familial hypercholesterolaemia: a machine learning study applied to electronic health encounter data. The Lancet Digital Health, Internet site: https://www.thelancet.com/journals/landig/article/PIIS2589-7500(19)30150-5/fulltext, last accessed November 2019.

[NECI]. 1997. Takeshi Murakami (0) vs. Logistello (6), Internet site: https://skatgame.net/mburo/event.html, last accessed November 2019.

Newell, A. and Shaw, J. C. 1956. The logic theory machine. Transactions on Information Theory, IT-2, No. 3, September.

Newell, A., Shaw, J. C. and Simon, H. A. 1959. Report on a general problem-solving program. Proceedings of the International Conference on Information Processing. Available online: http://bitsavers.informatik.uni-stuttgart.de/pdf/rand/ipl/P-1584_Report_On_A_General_Problem-Solving_Program_Feb59.pdf, last accessed November 2019.

Newell, A. and Simon, H. 1976. Computer science as empirical inquiry: symbols and search. Communications of the ACM, 19: 3.

Newell, A. 1982. The knowledge level. Artificial Intelligence, 18(1): 87–127.

Newell, A. 1990. Unified Theories of Cognition. Harvard University Press.

[NHTSA]. 2019. The road to full automation. National Highway Traffic Safety Administration Internet site: https://www.nhtsa.gov/technology-innovation/automated-vehicles-safety#topic-road-self-driving, last accessed November 2019.

Nkambou, R., Mizoguchi, R. and Bourdeau, J. 2010. Advances in Intelligent Tutoring Systems. Heidelberg: Springer.

Noroozi, F., Corneanu, C., Kaminska, D., Sapinski, T., Escalera, S. and Anbarjafari, G. 2018. Survey on emotional body gesture recognition. Journal of IEEE Transactions on Affective Computing. Available online: https://arxiv.org/pdf/1801.07481.pdf, last accessed December 2019.

Nwana, H. S. 1990. Intelligent tutoring systems: An overview. Artificial Intelligence Review, 4(4): 251–277.

Nwana, H. S. 1996. Software agents: An overview. Knowledge Engineering Review, 11(3): 1–40, Sept, Cambridge University Press.

Nyquist, H. 1924. Certain factors affecting telegraph speed. Bell System Technical Journal, 3: 324–346.

Nyquist, H. 1928. Certain topics in telegraph transmission theory. Trans. AIEE, 47: 617–644. Reprinted as classic paper in: Proc. IEEE, Vol. 90, No. 2, Feb 2002.

Obermeyer, Z. 2016. Predicting the future—big data, machine learning, and clinical medicine. New England Journal of Medicine, 375(13): 1216–1219, September.

Ohlheiser, A. 2016. Trolls turned Tay, Microsoft's fun millennial AI bot, into a genocidal maniac, 25 March 2016. Available online: https://www.washingtonpost.com/news/the-intersect/wp/2016/03/24/the-internet-turned-tay-microsofts-fun-millennial-ai-bot-into-a-genocidal-maniac/, last accessed October 2019.

Ornes, S. 2019. Playing hide-and-seek, machines invent new tools. Quanta Magazine. Available online: https://www.quantamagazine.org/artificial-intelligence-discovers-tool-use-in-hide-and-seek-games-20191118/, last accessed November 2019.

Pease, A. and Pease, B. 2004. The Definitive Book of Body Language. Bantam Books, New York.

Potter, L. E., Araullo, J. and Carter, L. 2013. The Leap Motion controller: a view on sign language. pp. 175–178. *In*: Haifeng Shen, Ross Smi+C10th, Jeni Paay, Paul Calder and Theodor Wyeld (eds.). Proceedings of the 25th Australian Computer-Human Interaction Conference: Augmentation, Application, Innovation, Collaboration (OzCHI '13). ACM, New York, NY, USA.

Pressey, S. L. 1926. A simple apparatus which gives tests and scores and teaches. School and Society, 23(586): 373–376.

Pressey, S. L. 1927. A machine for automatic teaching of drill material. School and Society, 25(645): 549–552.

Radford, A., Wu, J., Child, R., Luan, D., Amodei, D. and Sutskever, I. 2019. Language models are unsupervised multitask learners. OpenAI Internet site located at: https://cdn.openai.com/better-language-models/language_models_are_unsupervised_multitask_learners.pdf, last accessed December 2019.

Randhavane, T., Bhattacharya, T., Kapsaskis, K., Gray, K., Bera, A. and Manocha, D. 2019. Identifying emotions from walking using affective and deep features. University of North Carolina Chapel Hill Internet site located at: http://gamma.cs.unc.edu/GAIT/files/Emotion_LSTM.pdf, last accessed December 2019.

Raphael, R. 2019. Fitness startup Mirror claims it's 'building the next iPhone'. FastCompany Internet site located at: https://www.fastcompany.com/90434763/fitness-startup-mirror-has-big-plans-including-telemedicine-were-building-the-next-iphone, last accessed December 2019.

Raskar, R., Welch, G. and Fuchs, H. 1998. Spatially augmented reality. Proceedings of the First International Workshop on Augmented Reality, San Francisco, November. Available online: https://pdfs.semanticscholar.org/7010/3d40f4 05d3a86ac2973d1d8d89576e755b86.pdf, last accessed December 2019.

Rasool, A. 2019. What is Augmented Reality in Smartphones? Datafloo Internet site located at: https://datafloq.com/read/what-is-augmented-reality-in-smartphones/5785, last accessed December 2019.

Reece-Hedberg, S. 2002. DART: Revolutionizing logistics planning. IEEE Intelligent Systems, 17(3): 81–83.

[Riken]. 2015. The strong robot with the gentle touch. Riken Internet site located at: https://www.riken.jp/en/news_pubs/research_news/2015/20150223_2/, last accessed December 2019.

Rogers, E. M. 2003. Diffusion of Innovations (5th ed.). New York, NY: Free Press.

Rometty, G. 2016. CES 2016 Keynote Address, YouTube video available at https://www.youtube.com/watch?v=VEq-W-4iLYU.

Rosenblatt, F. 1958. The perceptron: A probabilistic model for information storage and organization in the brain. Psychological Review, 65(6): 386–408.

Rosenschein, J. S. and Genesereth, M. R. 1988. Deals among rational agents. Readings in Distributed Artificial Intelligence. Morgan Kaufman.

[ROSS]. 2019. We're building technology to help everyone obtain the best possible legal outcomes. ROSS Intelligence Internet site located at: https://rossintelligence.com/about.html, last accessed December 2019.

Rumelhart, D. E., Hinton, G. E. and Williams, R. J. 1986. Learning representations by back-propagating errors. Nature, 323: 533–536.

Russell, S. and Norvig, P. 2009. Artificial Intelligence: A Modern Approach. 3rd Edition, Pearson.

[Ruuh]. 2018. Happy Birthday, Ruuh! Microsoft's AI-powered desi chatbot turns one. Microsoft Internet site located at: https://www.microsoft.com/en-in/campaign/artificial-intelligence/ruuh-ai-chatbot-one-year-anniversary.aspx, last accessed December 2019.

Samuel, A. L. 1959. Some studies in machine learning using the game of checkers. IBM Journal of Research and Development, 44: 206–226.

Sandoiu, A. 2019. Artificial intelligence better than humans at spotting lung cancer. Medical News Today Newsletter, May 20. Available online: https://www.medicalnewstoday.com/articles/325223.php#1, last accessed November 2019.

Sarikaya, R. 2017. The technology behind personal digital assistants: An overview of the system architecture and key components. IEEE Signal Processing Magazine, 34(1): 67–81.

Saygin, A. P., Cicekli, I. and Akman, A. P. 2000. Turing test: 50 years later. Minds and Machines, Vol. 10.

Schermer, B. W. 2007. Software Agents, Surveillance, and the Right to Privacy: A Legislative Framework for Agent-Enabled Surveillance, Leiden, retrieved October 9,2018 from https://openaccess.leidenuniv.nl/bitstream/handle/1887/11951/Thesis.pdf.

Schmidhuber, J. 2019. History of Robotic Cars. Available online: http://people.idsia.ch/~juergen/robotcars.html, last accessed November 2019.

Schumpeter, J. 1942. Capitalism, Socialism and Democracy. Harper and Bros.

Schutt, K. T., Gastegger, M., Tkatchenko, A., Muller, K. R. and Maurer, R. J. 2019. Unifying machine learning and quantum chemistry with a deep neural network for molecular wavefunctions. Nature Communications, (10): 5024. Available online at: https://www.nature.com/articles/s41467-019-12875-2#citeas, last accessed December 2019.

[ScriptBook]. 2019. To greenlight or not to greenlight, that is the question. ScriptBook Internet site located at: https://www.scriptbook.io/#!/scriptbook/proof, last accessed December 2019.

Searle, J. 1980. Minds, brains and programs. Behavioral and Brain Sciences, 3(3): 417–457.

Sejnowski, T. and Rosenberg, C. 1987. Parallel Networks that Learn to Pronounce English Text Complex Systems 1: 145–168. Available online: http://citeseerx.ist.psu.edu/viewdoc/download?doi=10.1.1.154.7012&rep=rep1&type=pdf, last accessed November 2019.

Shannon, C. E. 1948. A mathematical theory of communication. Bell System Technical Journal, 27(3): 379–423.

Shapiro, G. P. 1990. Knowledge discovery in real databases: A report on the IJCAL-89 workshop. AI Magazine, 11(5): 68–70.

Shivley, R. J., Bandt, S. L. Lachter, J., Matessa, M., Sadller, G. and Battise, V. 2016. Application of Human Autonomy Teaming (HAT) Patterns to Reduce Crew Operations (RCO), National Aeronautics and Space Administration, retrieved September 17, 2018 from https://ntrs.nasa.gov/archive/nasa/casi.ntrs.nasa.gov/20160006634.pdf.

Shivley, R. J., Bandt, S. L. Lachter, J., Matessa, M., Sadller, G., Battise, V. and Johnson, W. 2018. Why Human Autonomy Teaming. ResearchGate, retrieved October 9, 2018, from https://www.researchgate.net/publication/318182279_Why_Human-Autonomy_Teaming.

Shum, H., He, X. and Li, D. 2018. From Eliza to XiaoIce: challenges and opportunities with social chatbots. Frontiers of Information Technology & Electronic Engineering, 19: 10–26.

Silberschatz, A. and Tuzhilin, A. 1995. On subjective measures of interestingness in knowledge discovery. KDD-95 Proceedings, Association for the Advancement of Artificial Intelligence. Available online: https://www.aaai.org/Papers/KDD/1995/KDD95-032.pdf, last accessed November 2019.

Silver, D., Huang, A., Maddison, C. J., Guez, A., Sifre, L., Driessche, G., Schrittwieser, J., Antonoglou, I., Panneershelvam, V., Lanctot, M., Dieleman, S., Grewe, D., Nham, J., Kalchbrenner, N., Sutskever, I., Lillicrap, T., Leach, M., Kavukcuoglu, K., Graepel, T. and Hassabis, D. 2016. Mastering the game of Go with deep neural networks and tree search. Nature, 529: 484–489.

Silver, D., Schrittwieser, J., Simonyan, K., Antonoglou, J., Huang, A., Guez, A., Hubert, T., Baker, L., Lai, M., Bolton, A., Chen, Y., Lillicrap, T., Hui, F., Sifre, L., Driessche, G., Graepel, T. and Hassabis, D. 2017. Mastering the game of Go with deep neural networks and tree search. Nature, 550: 354–359.

Simon, H. A. 1956. Rational choice and the structure of the environment. Psychological Review, 63(2): 129–138.

Simon, H. A. and Gilmartin, K. J. 1973. A simulation of memory for chess positions. Cognitive Psychology, 5: 29–46.

Simonite, T. 2014. Facebook creates software that matches faces almost as well as you do. MIT Technology Review. Available online: https://www.technologyreview.com/s/525586/facebook-creates-software-that-matches-faces-almost-as-well-as-you-do/, last accessed November 2019.

Sinclair, M. 2019. Why the self-help industry is dominating the U.S.: A brief history of self-improvement. Medium Internet site located at: https://medium.com/s/story/no-please-help-yourself-981058f3b7cf, last accessed December 2019.

Singh, S., Okun, A. and Jackson, A. 2017. Learning to play Go from scratch. Nature, 550: 336–337.

Skinner, B. F. 1958. Teaching machines. Science, Vol. 128, Iss. 3330, October.

Skinner, B. F. 1965. The technology of teaching. Proceedings of the Royal Society, Vol. 162.

Slagle, J. R. 1961. A heuristic program that solves symbolic integration problems in freshman calculus: symbolic automatic integrator (SAINT). Massachusetts Institute of Technology. Available online: https://dspace.mit.edu/bitstream/handle/1721.1/11997/31225400-MIT.pdf?sequence=2, last accessed November 2019.

Smith, D. 2019. Amazon Echo Loop—everything we know about Amazon's wacky new smart ring. CNet Internet site located at: https://www.cnet.com/how-to/amazon-echo-loop-everything-we-know-about-amazons-wacky-new-smart-ring/, last accessed December 2019.

Spice, B. 2017. A computer that reads body language. Carnegie Mellon Internet site located at: https://www.cs.cmu.edu/news/computer-reads-body-language, last accessed December 2019.

Steels, L. 1990. Components of expertise. AI Magazine, Vol. 11, No. 2.

Steptoe, A., Shankar, A., Demakakos, P. and Wardle, J. 2013. Social isolation, loneliness, and all-cause mortality in older men and women. Proceedings of the National Academy of Sciences of the United States of America (PNAS), Vol. 110, No. 15, April 9.

[StoryFit]. 2019. Make more informed decisions from script to sales. StoryFit Internet site located at: https://storyfit.com/storyfit-for-studios/, last accessed December 2019.

Sun, R. 2002. Duality of the Mind: A Bottom-up Approach Toward Cognition. Mahwah, NJ: Lawrence Erlbaum Associates.

Takahashi, D. 2015. Elemental's smart connected toy CogniToys taps IBM's Watson supercomputer for its brains, Venture Beat, 23 February 2015. Available online: https://venturebeat.com/2015/02/23/elementals-smart-connected-toy-cognitoys-taps-ibms-watson-supercomputer-for-its-brains/, last accessed September 2019.

Tesauro, G. 1995. Temporal difference learning and TD-Gammon. Communications of the ACM, Vol. 38, No. 3.

[Therapy]. 2019. Therapy. Psychology Today Internet site located at: https://www.psychologytoday.com/us/basics/therapy, last accessed December 2019.

Thompson, C. 2002. Approximating Life, The New York Times, 7 July 2002. Available online: https://www.nytimes.com/2002/07/07/magazine/approximating-life.html, last accessed September 2019.

Tran, K. and Ulissi, Z. W. 2018. Active learning across intermetallics to guide discovery of electrocatalysts for CO_2 reduction and H_2 evolution. Nat. Catal., 1.

Tshitoyan, V., Dagdelen, J., Weston, L., Dunn, A., Rong, Z., Kononova, O., Persson, K. A., Ceder, C. and Jain, A. 2019. Unsupervised word embeddings capture latent knowledge from materials science literature. Nature, Vol. 571, July.

Turing, A. M. 1936. On computable numbers, with an application to the Entscheidungsproblem. Proceedings of the London Mathematical Society. 2(published 1937). 42: 230–265.

Turing, A. M. 1947. Lecture to the London Mathematical Society on 20 February 1947. Available online: http://www.vordenker.de/downloads/turing-vorlesung.pdf, last accessed November 2019.

Turing, A. M. 1950. Computing machinery and intelligence. Mind, Vol. LIX, Issue 236, October 1950.

Ulissi, Z. W., Tang, M. T., Xiao, J., Liu, X., Torelli, D. A., Karamad, M., Cummins, K., Hahn, C., Lewis, N. S., Jaramillo, T. F., Chan, K. and Nørskov, J. K. 2017. Machine-learning methods enable exhaustive searches for active bimetallic facets and reveal active site motifs for CO_2 reduction. ACS Catal., 7: 6600–6608.

Upbin, B. 2013. IBM's Watson gets its first piece of business in healthcare. Forbes, Internet page located at: http://www.forbes.com/sites/bruceupbin/2013/02/08/ibms-watson-gets-its-first-piece-of-business-in-healthcare/#780177ca44b1 and last accessed March 2016.

[USPTO]. 2019. Request for comments on intellectual property protection for artificial intelligence innovation. Federal Register, Vol. 84, No. 210, Wednesday, October 30. Available online: https://www.govinfo.gov/content/pkg/FR-2019-10-30/pdf/2019-23638.pdf, last accessed November 2019.

Wallace, R. 2009. The anatomy of A.L.I.C.E. In The Parsing of the Turing Test, pp. 181–210.

Warwick, K. and Shah, H. 2016. The importance of a human viewpoint on computer natural language capabilities: A Turing test perspective. AI & Society, 31(2): 207–221.

[WashPostPR]. 2016. The Washington Post experiments with automated storytelling to help power 2016 Rio Olympics coverage. Washington Post Internet site located at: https://www.washingtonpost.com/pr/wp/2016/08/05/the-washington-post-experiments-with-automated-storytelling-to-help-power-2016-rio-olympics-coverage/, last accessed December 2019.

Watkins, C. J. C. H. 1989. Learning from Delayed Rewards. Ph.D. Thesis, University of Cambridge, Cambridge. Available online: https://www.cs.rhul.ac.uk/home/chrisw/new_thesis.pdf, last accessed November 2019.

Weiner, N. 1948. Cybernetics: Or Control and Communication in the Animal and the Machine (Hermann & Cie) & Camb. Mass. (MIT Press) 2nd Revised Ed. 1961.

Weiser, M. 1991. The computer for the 21st century, scientific American special issue on communications. Computers, and Networks, September.

Weizenbaum, J. 1966. ELIZA—a computer program for the study of natural language communication between man and machine. Communications of the ACM, 9(1): 36–45.

Weizenbaum, J. 1976. Computer Power and Human Reason: From Judgement to Calculation, New York: W.H. Freeman and Company.

[WHO]. 2017. Mental health of older adults. World Health Organization Internet site located at https://www.who.int/news-room/fact-sheets/detail/mental-health-of-older-adults, last accessed December 2019.

Winograd, T. 1971. Procedures as a representation for data in a computer program for understanding natural language's dissertation. MIT AI Technical Report 235, also Journal Cognitive Psychology Vol. 3, No. 1.

Winograd, T. 1972. Understanding Natural Language. Academic Press.

Winston, P. H. 1970. Learning structural descriptions from examples. Massachusetts Institute of Technology. Dept. of Electrical Engineering. Thesis. Available online: https://dspace.mit.edu/handle/1721.1/13800, last accessed November 2019.

Winston, P. H. 1977. Artificial Intelligence. Addison-Wesley Pub. Co.

Winston, P. H. 2011. The strong story hypothesis and the directed perception hypothesis. *In*: Pat Langley (ed.). AAAI Fall Symposium Series (2011), Association for the Advancement of Artificial Intelligence.

Wladawsky-Berger, I. 2013. The era of augmented cognition. The Wall Street Journal: CIO Report, Internet page located at http://blogs.wsj.com/cio/2013/06/28/the-era-of-augmented-cognition/, last accessed May 2019.

Woodie, A. 2014. Inside Sibyl, Google's massively parallel machine learning platform, Datanami. Tabor Communications. Available online: https://www.datanami.com/2014/07/17/inside-sibyl-googles-massively-parallel-machine-learning-platform/, last accessed November 2019.

Zhang, L. 2017. From machine learning to deep learning: progress in machine intelligence for rational drug discovery. Drug Discovery Today, 22(11): 1680–1685, November.

Zhou, L., Gao, J., Li, D. and Shum, H. 2018. The design and implementation of XiaoIce, an empathetic social chatbot. ArXiv, vol. abs/1812.08989.

Zimmerman, T. G., Lanier, J., Blanchard, C., Bryson, S. and Harvill, Y. 1987. A hand gesture interface device. ACM SIGCHI Bulletin, 18(4): 189–192.

[Zo]. 2019. Let's Talk About Zo, Microsoft Internet site located at: https://www.zo.ai/, last accessed December 2019.

Zolfagharifard, E. 2015. Smart toilet than can analyse your pee and an app that detects depression: Japanese expo reveals strange health gadgets. Daily Mail Internet site located at: https://www.dailymail.co.uk/sciencetech/article-3306193/Smart-toilet-analyse-PEE-app-detects-depression-Japanese-expo-reveals-strange-health-gadgets.html, last accessed December 2019.

Index